EAI/Springer Innovations in Communication and Computing

Series Editor
Imrich Chlamtac, European Alliance for Innovation, Ghent, Belgium

Editor's Note

The impact of information technologies is creating a new world yet not fully understood. The extent and speed of economic, life style and social changes already perceived in everyday life is hard to estimate without understanding the technological driving forces behind it. This series presents contributed volumes featuring the latest research and development in the various information engineering technologies that play a key role in this process.

The range of topics, focusing primarily on communications and computing engineering include, but are not limited to, wireless networks; mobile communication; design and learning; gaming; interaction; e-health and pervasive healthcare; energy management; smart grids; internet of things; cognitive radio networks; computation; cloud computing; ubiquitous connectivity, and in mode general smart living, smart cities, Internet of Things and more. The series publishes a combination of expanded papers selected from hosted and sponsored European Alliance for Innovation (EAI) conferences that present cutting edge, global research as well as provide new perspectives on traditional related engineering fields. This content, complemented with open calls for contribution of book titles and individual chapters, together maintain Springer's and EAI's high standards of academic excellence. The audience for the books consists of researchers, industry professionals, advanced level students as well as practitioners in related fields of activity include information and communication specialists, security experts, economists, urban planners, doctors, and in general representatives in all those walks of life affected ad contributing to the information revolution.

About EAI

EAI is a grassroots member organization initiated through cooperation between businesses, public, private and government organizations to address the global challenges of Europe's future competitiveness and link the European Research community with its counterparts around the globe. EAI reaches out to hundreds of thousands of individual subscribers on all continents and collaborates with an institutional member base including Fortune 500 companies, government organizations, and educational institutions, provide a free research and innovation platform.

Through its open free membership model EAI promotes a new research and innovation culture based on collaboration, connectivity and recognition of excellence by community.

More information about this series at http://www.springer.com/series/15427

Dagmar Cagáňová • Natália Horňáková
Andy Pusca • Pedro Filipe Cunha
Editors

Advances in Industrial Internet of Things, Engineering and Management

Springer

EAI
RESEARCH MEETS INNOVATION

Editors
Dagmar Cagáňová
Faculty of Materials Science
and Technology in Trnava
Slovak University of Technology
in Bratislava
Trnava, Slovakia

Natália Horňáková
Faculty of Materials Science
and Technology in Trnava
Slovak University of Technology
Trnava, Slovakia

Andy Pusca
Danubius University
Galati, Romania

Pedro Filipe Cunha
Instituto Politecnico de Setubal
Setubal, Portugal

ISSN 2522-8595 ISSN 2522-8609 (electronic)
EAI/Springer Innovations in Communication and Computing
ISBN 978-3-030-69704-4 ISBN 978-3-030-69705-1 (eBook)
https://doi.org/10.1007/978-3-030-69705-1

This Springer imprint is published by the registered company Springer Nature Switzerland AG
The registered company address is: Gewerbestrasse 11, 6330 Cham, Switzerland

Preface

This scientific monograph is a comprehensive work of the authors and editors who are experts in the area of industrial internet of things, engineering and management. The authors of the scientific chapters are academicians, practitioners and researchers from countries like Finland, Ukraine, Romania and Slovakia.

Although the scientific monograph focuses on the experience and deep knowledge of its chapters' authors, the international readers may find it useful also for individual approach to the common topic.

The scientific monograph consists of the following 11 chapters:

Chapter 1 proposes an algorithm that automatically determines the test data amount, depending on the bandwidth of the network channel. The measurement procedure eliminates operator error. According to the results of the research, the authors revealed the hours of maximum and minimum load and load extrema, which is characterized by the fact that the number of subscribers using the cellular network for voice calls or data transmission is the maximum.

Chapter 2 examines the impact of various levels of ICT deployment on the driving time of an emergency response vehicle under specific traffic conditions near a site where a large cultural or sports event takes place. The driving time of the emergency response vehicle is evaluated by using realistic simulation of both traffic scenarios and communication technologies deployment. The impact of communication on emergency vehicle's driving time is evaluated for varying penetration of connected vehicles.

Chapter 3 puts the focus on the specifics of the retail network of three regional towns in the Slovak Republic—Nitra, Trnava and Žilina—and answers the following question: What value of the Population to the Admissible Floor Space (PAFS) indicator will cause saturation of the retail network?

Chapter 4 is oriented toward developing a car accident detecting system using data obtained from video recording streams applying machine learning algorithms. The system can be successfully applied in various industries and helps react to car accidents faster and control traffic flows efficiently.

Chapter 5 deals with the design and implementation of a new-generation platform for the smart city ecosystem that offers valuable information on multiple urban dimensions with direct impact on a large spectrum of users with different interests. The software Citizen Incident Reporting allows citizens to effectively participate in the local governance by documenting their concerns and sending reports to the government order management services.

Chapter 6 deals with the research and analysis of the perception of creativity and its importance for managers of different generation groups in the Slovak industrial enterprises. Creative thinking in the industrial practice is important from the perspective of the enterprise prosperity interests and innovation. Managers use creativity in everyday activities such as planning or solving problems.

Chapter 7 deals with the aspects of the development of mobile communication systems, taking into account promising technologies and ITU recommendations. The basic principles of construction of mobile telecommunications networks are given, taking into account the potential implementation of any scenario for the implementation of virtual technical functions with established requirements for a given quality of service.

Chapter 8 presents a framework for energy consumption prediction in a household. The framework collects real data from a residential house by a collection of sensors and then prepossesses the data. After that, it utilizes two well-known prediction models, multilayer perceptron (MLP) and K-nearest neighbour (K-NN), in order to predict energy consumption at the current time. Furthermore, it employs long short-term memory (LSTM) as one of the common recurrent neural networks for forecasting the next hour energy consumption.

Chapter 9 deals with the analysis (using ANSYS software) of three different materials that can be used in the manufacture of the electric car charger shell in terms of stresses and deformations occurring under wind action.

Chapter 10 considers one of the essential characteristics of the pumps as the head-capacity curve. The shape of this characteristic depends on the type of pump, its design features and other parameters. Nowadays, for obtaining the head-capacity curve, computer simulation is widely used through computer programs which are based on the finite volume method. The most popular program for this aim is ANSYS CFX. The results of numerical simulation from this program have a reasonable correlation with the experimental results. Pump design has an inducer and booster, which are considered in this chapter. For optimization of their design by using the aforementioned program, the parametric geometries are necessary. The method for making them is described in this chapter.

Chapter 11 presents a management decision-making approach of logistics aspects in port during interactions between two kinds of transport. The described methodology is developed according to the Industry 4.0 concept. The methodology considers two restrictions related to technical peculiarities of the internal port infrastructure and technological aspects of the interaction between two kinds of transport. The hybrid neuro-fuzzy ANFIS system is used as a predictor.

Finally, the editors would like to express their sincere thanks to the authors of the chapters for contributing their outstanding knowledge, experience and latest research results towards the creation of this scientific monograph.

Trnava, Slovakia

Dagmar Cagáňová
Natália Horňáková

Contents

Basic Stations Work Optimization in Cellular Communication Network

Anatolii Tkachuk, Valentyn Zablotskyi, Oleg Zabolotnyi, Dagmar Cagáňová, and Nataliia Yakymchuk

1 Introduction

Now, mainly such companies as Kyivstar, Vodafone Ukraine, Lifecell and Intertelecom provide 3G and 4G cellular services in Ukraine.

The launch of 1.8 GHz 4G networks in Ukraine has allowed Vodafone to significantly increase the coverage of high-speed Internet. In comparison with the previous period of operation, mobile Internet in the 2.6 GHz was available only in places with the largest concentration of Internet users. According to the operator, 4G service from Vodafone is available for 11.7 million Ukrainians. The geography of coverage of the fourth-generation network includes the entire territory of Ukraine (Fig. 1).

After the launch of 4G, Vodafone subscribers have the opportunity to actively use mobile Internet and 4G-based services (HD TV streaming, online games), actively use cloud services. Thanks to the introduction of new communication technologies, new directions have also become possible for use: telemedicine, smart-management of energy and agriculture, and electronic mobile wallets.

Because Kyivstar is one of the oldest operators in the mobile market of Ukraine, the company was able to increase the largest base of subscribers (approximately 27 million), as well as create a wide map of mobile Internet coverage (Fig. 2). Today it is a leader in the growth rate of new subscribers, as well as in the amount of traffic

A. Tkachuk (✉) · V. Zablotskyi · O. Zabolotnyi · N. Yakymchuk
Technological Faculty, Lutsk National Technical University, Lutsk, Ukraine
e-mail: a.tkachuk@lntu.edu.ua; v.zablotsky@lntu.edu.ua; m.yakymchuk@lntu.edu.ua

D. Cagáňová
Faculty of Materials Science and Technology in Trnava, Slovak University of Technology in Bratislava, Trnava, Slovakia
e-mail: dagmar.caganova@stuba.sk

D. Cagáňová et al. (eds.), *Advances in Industrial Internet of Things, Engineering and Management*, EAI/Springer Innovations in Communication and Computing,
https://doi.org/10.1007/978-3-030-69705-1_1

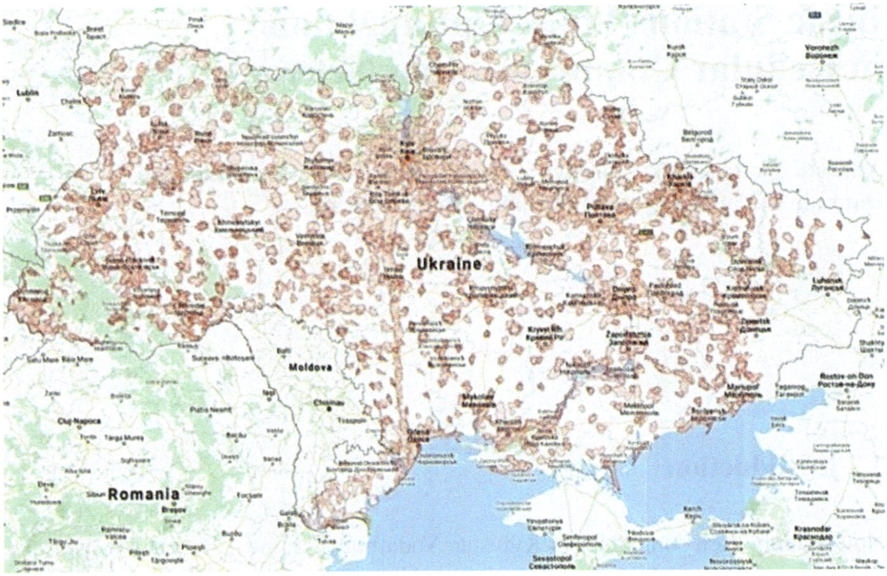

Fig. 1 4G zone coverage of the territory of Ukraine by "Vodafone". (Source: [1])

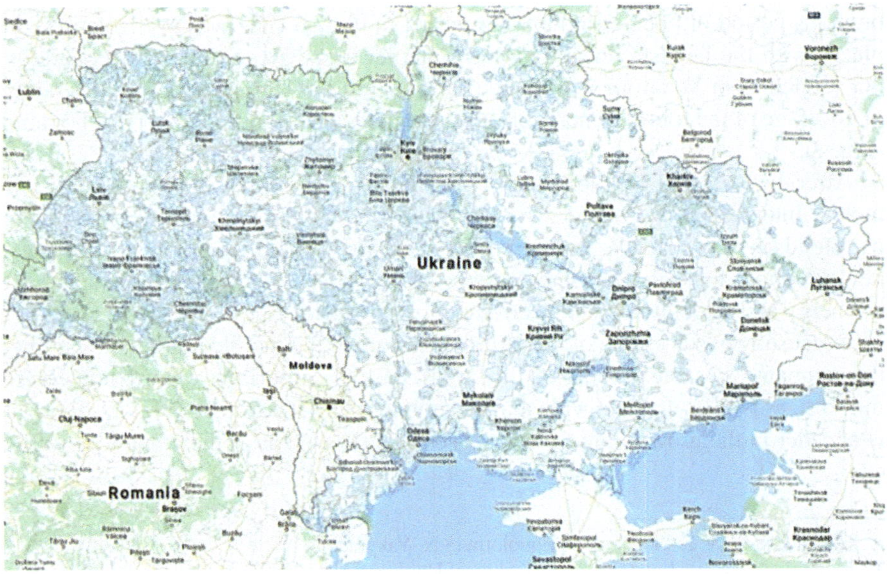

Fig. 2 4G zone coverage of the territory of Ukraine by "Kyivstar". (Source: [1])

consumed. The operator Kyivstar 4G coverage is developing very rapidly. At the auction of the fourth-generation frequencies, Kyivstar was able to get 30 MHz in the 2600 range and 70 MHz in the 1800 range, which is the highest indicator among competitors.

The main advantages of the new standard are a multiple increase in Internet speed (more than 15 times), improved ping, and no drop in speed while driving.

A very important indicator when choosing an operator is the quality of reception not only in large cities, but also outside them. The most common in this case is 3G Internet.

One of the operators providing wireless Internet services in Ukraine is the company Intertelecom. 3G Internet of this operator covers a significant part of the country. This company is the only one in the country, which, thanks to EV-DO technology, gives the opportunity to take advantage of the connection with a unique Speed B+ capable of transmitting data at a speed of about 14.7 Mbps.

On the territory of Ukraine, Intertelecom is a leader in providing 3G Internet connection; the coverage map covers at least 14,000 settlements and all regional centres (Fig. 3). Over 16 years of operation, the map of the location of Intertelecom base stations has become the most extensive and dense thanks to the use of fibre-optic lines and SDH multiplexer nodes. The base stations are located in such a way that high-speed wireless communication is provided to all major roads and railways, also in the area of action falls and most of the countryside. The coverage area is more than 85% of the country.

The Lifecell mobile operator uses a 4G network in the 1800 MHz band. Today, the third- and fourth-generation technology from Lifecell covers 2252 settlements of the country and is available to more than 22 million users (Fig. 4).

As seen in Figs. 1, 2, 3 and 4, mobile operators provide communication services in the territory of the majority of settlements of Ukraine. It is also seen that there

Fig. 3 Zone 3G and 4G coverage of the territory of Ukraine by "Intertelecom". (Source: [1])

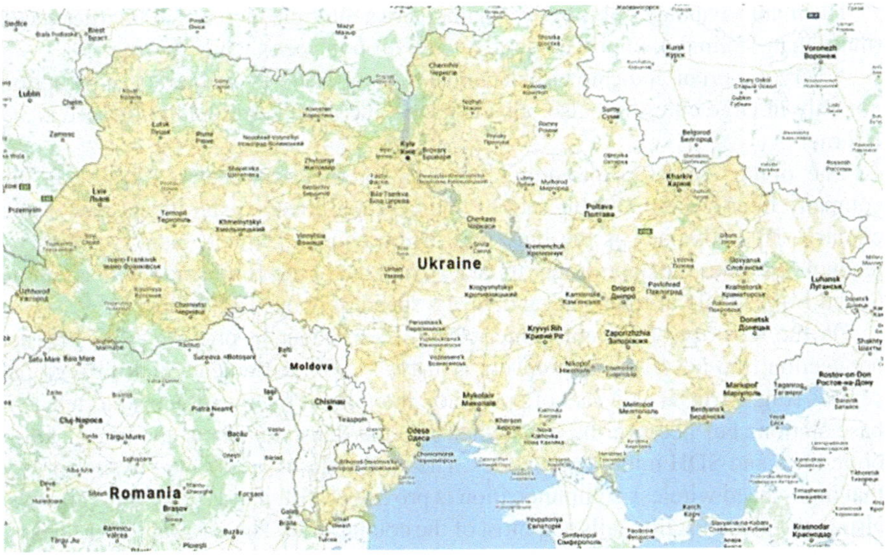

Fig. 4 3G and 4G zone coverage of Ukraine by "Lifecell". (Source: [1])

is a trend of transition of subscribers from active use of voice data to active use of Internet services. In this regard, the requirements to the criteria of information transmission quality, such as high bit rates, stability of channel bandwidth, and network capacity, are increased.

Currently, two generations of cellular communication coexist and function simultaneously. The third generation of WCDMA is the most common, as its development and implementation began much earlier than the fourth generation of LTE. WCDMA is used to transmit voice calls because third-generation subscriber terminals are the most common. Since such subscriber terminals are cheaper to manufacture and easier to maintain, the third generation of subscriber terminals will exist in the market of communication services for a certain period. In addition, the re-equipment of communication systems, as well as the replacement of subscriber terminals, will be an economically difficult task.

Simultaneously with WCDMA, the fourth-generation LTE system is deployed. These systems can work simultaneously without interfering with each other, because they have a different frequency range. The fourth generation of cellular communication systems has a much larger bandwidth of the radio channel, and therefore has a much higher data rate. The fourth generation of LTE communication is mainly used for broadband Internet access, as well as for video calls and viewing a wide range of multimedia information in streaming mode.

Thus, two generations of communication systems complement each other, simultaneously operating on the same section of the network.

Quality control of all services provided, both voice and the mobile operator should perform data transmission for both generations.

Operators need to maintain the quality of services at a high level, so there is a need to test periodically cellular networks.

One of the important parameters that affect the quality of communication is the bit rate of the Internet connection. The problem of providing high-speed Internet connection is exacerbated by the growth of data transmitted through the Internet, which is now very actively used to view multimedia information in streaming mode (TV movies, online channels, mobile TV, etc.), as well as to visit social networks, where information pages are overloaded with graphic information. For comfortable work on the Internet now you need speed 1.5–2 Mbps. Therefore, the requirements to the quality of the data transmission channel, as well as to the stability of technical and operational characteristics of cellular networks, are increased.

Therefore, an important task is to monitor the speed of your Internet connection, which currently uses a variety of Internet resources that allows you to quickly determine the connection speed, but have several significant limitations such as: the limited number of queries per day, the impossibility of testing the speed of your Internet connection without user interaction (automated testing), and the inability to arbitrarily change the size of the test package. In this regard, there is a need to develop a test system independent of any external services.

The procedure for testing the communication channel is as follows. A test file of a certain size is generated on the remote server and then the generated file is transferred to the device under test. The data rate is defined as the amount of data transferred over a period. Typically, this interval is 1 s, but sometimes it is necessary to determine the average data rate per hour or per day. During testing, the operator selects a broadband connection, and the speed test server generates a large test file, which is necessary to determine accurately the data rate [2–4]. However, for the low speed, testing takes longer than planned, which can lead to errors in determining the speed of the Internet connection or "hang" service [5–7].

2 Analysis of Recent Research and Publications

In order to avoid such situations, it is proposed to use an adaptive algorithm for measuring the speed by changing the size of the downloaded file. Speed measurements will be performed in the following way: first, a small file is transferred and the bit rate is determined relative to the control time interval. If the test file is transferred immediately, the second step is to repeat the procedure with a larger file, and so on until the results of two consecutive measurements are of the same order, then the results are averaged and displayed to the operator as the results of the measurement [8–10]. In order to improve the accuracy of measuring the bit rate of the Internet connection, as well as reducing the total testing time, the following volumes of test files are proposed (Table 1).

The authors of the paper recommend to use a 10 kB test file size to test the efficiency of the communication channel. With any quality of communication channel, such a small amount of information should be transmitted to the test server.

Table 1 Recommended file sizes for Internet connection speed testing (own measurements)

Test file size	Purpose of this test
10 kB	Test line check
200 kB	Testing EDGE (GPRS) connections
800 kB	Testing EDGE (GPRS) connections (improved accuracy)
1.6 MB	Testing HSDPA/HSUPA connection up to 3.6 Mbps
8 MB	Testing HSDPA/HSUPA connection up to 14.4 Mbps
16 MB	Testing HSDPA/HSUPA connection up to 14.4 MBps (increased accuracy)

Then it can be estimated the time for which the data were transferred. If the time interval equals to 2 is exceeded to 4, it can be considered a test of a complete channel, as studied the data channel is the low speed.

A data volume of 200 kB is used to test the data link that is generated when 2G networks are used at 384 kbps EDGE (GPRS) speeds. This file size allows to get an upper estimation of the Internet connection speed. To determine the speed of an Internet connection more accurately, a larger amount of data is required in order to measure the speed of multiple passes. To do this, the authors of the paper recommend to use a test file with 800 kB of data. Testing the third-generation networks requires much more data to determine adequately the bit rate. It is recommended to start testing in WCDMA networks with a test file with 1.6 MB of information. At an initial data transmission speed is 3.8 Mbps file will be passed through 8–10 s. For later versions of the HSDPA standard, up to 14.4 Mbps data transfer rates are achieved using 15 channel gate codes and 16-QAM modulation. At these high data rates, it is recommended to use 8 MB of data and 16 MB of data to determine the data rate more accurately. After carrying out a test study of the first file, to select the second file, then if the file is transferred in the same short time, in several stages can be skipped and immediately selected a larger file, which will more accurately determine the bit rate of the Internet connection.

3 Material and Method

To ensure the high quality of services provided by the operator, it is necessary to constantly diagnose and monitor various parameters of cellular communication systems. According to ISO 9000-2001, the quality of services should be ensured at all stages of customer service. Let us consider the main parameters of indicators for voice calls:

- Ensuring the normative level of quality indicators.
- Restore communication in case of failures.
- Carrying out preventive measures for the operation of the operator's equipment.

One of the most important factors is the provision of access to the network in the zone of confident reception and ensuring the normative level of service quality indicators in the zone of confident reception; therefore, it is proposed to use the following methodology: when assessing the continuity of communication, calls that meet the requirements for three main parameters are evaluated.

1. Availability of communication is the share of successful calls from the total number of calls for the established connection with the subscriber, both mobile and fixed network. The share of successful P_y calls is estimated by measuring the number of successful and non-successful calls made by network subscribers in the direction of network subscribers and fixed network subscribers:

$$P_y = Q/N \tag{1}$$

where N—total number of control calls for all measurement sessions, Q—total number of successful control calls for all measurement sessions.

2. Continuity of communication—the proportion of calls that ended in disconnection of the established connection at the initiative of the subscriber. The proportion of calls with planned disconnection P_n is estimated by measuring the number of successful calls and calls with premature disconnection:

$$P_n = R/N \tag{2}$$

where R—the total number of control calls that ended with disconnection of the established connection at the initiative of the subscriber.

3. The share of calls that meet the standards for the quality of speech. The proportion of calls that meet the standards for the quality of speech is estimated by measuring the number of calls with satisfactory and unsatisfactory quality of speech.

$$R_y = N_{np}/N \tag{3}$$

where N_{np}—the total number of control calls that meet the standards for the quality of speech for all measurement sessions.

Thus, in order to determine the listed quality indicators, a sufficiently large number of voice calls must be made on the network segment. During the testing of the entire network coverage area, the total number of voice calls will be very large. An additional parameter to which high requirements are imposed is the speed of data transmission when using the Internet by the subscriber. For comfortable work on the Internet currently enough speed is 1.5–2 Mbps.

Consequently, the requirements to the quality of the data channel of cellular networks, as well as to the stability of the characteristics of the declared services, are becoming stricter. Therefore, regular testing and monitoring is necessary to ensure high quality of services.

The minimum load hours are the period in which the number of subscribers on the investigated site is minimal. Based on the daily bit rate monitoring data, it is possible to detect such clocks and reallocate resources to other sectors where

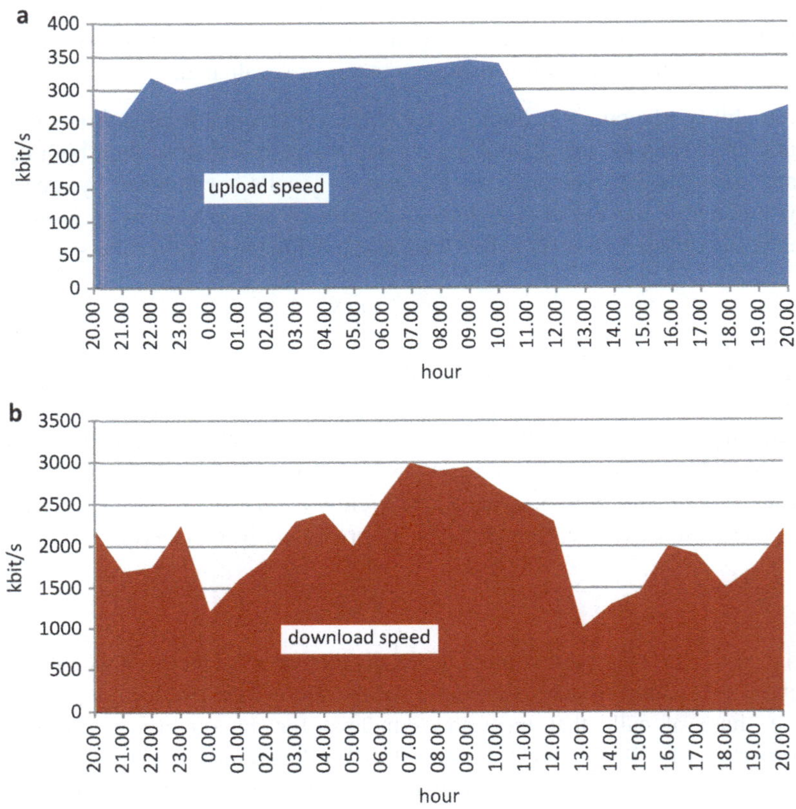

Fig. 5 Diagram of the daily change in the speed of the Internet connection for unloading (**a**) and loading data (**b**). (Source: Own measurements)

necessary. The diagram of daily change of Internet connection speed for data unloading (a) and loading (b) is presented in Fig. 5.

The graphs show that the minimum download speed (Fig. 5b) observed at about one o'clock in the day, it follows that this point corresponds to the maximum load on the network.

The minimum load on the network was recorded at seven in the morning; this experimentally determined maximum speed coincides with the data from the LAC controller. Data from the controller on the load factor of the base station in conventional units for the control period are shown in Fig. 6.

The operator to 300 kbps artificially limits the data upload speed. This is done to free up network resources. For daily monitoring of Internet connection speed in the cellular network, the described method is quite expensive, because a large amount of traffic is spent. In this example, about 22 GB of traffic was spent during testing. The average daily rate can be calculated as:

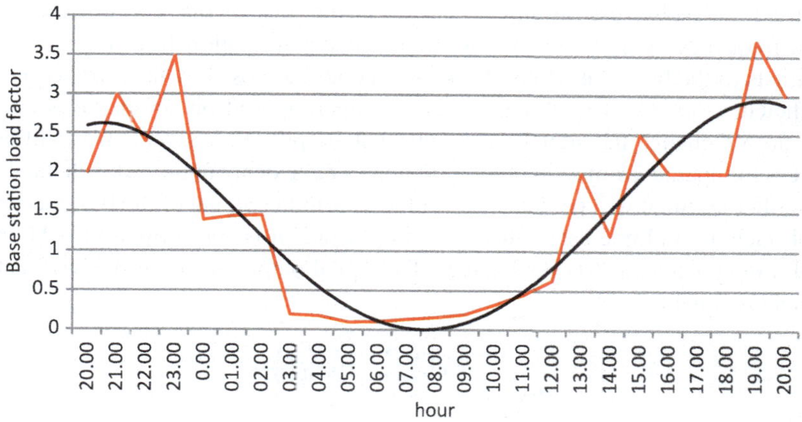

Fig. 6 Changing the load factor of the base station per day. (Source: Own measurements)

$$S_{CP} = \sum_{k=1}^{n} \frac{S_k}{n} \tag{4}$$

where S_{CP}—average daily bit rate, bit/s, S_k—bit rate of a single measurement bit/s, and n—number of measurements.

In the example shown in Fig. 5b, the average daily download speed when using an Internet connection is 1283 kbps. For daily network monitoring, the described method is ineffective, since a large flow of traffic occurs through continuous data exchange, which will take resources from users, and, accordingly, in the hours of maximum load will cause failures in different communication services.

4 Results

An alternative, passive method for determining the hours of maximum and minimum load, as well as the maximum and minimum speeds of the Internet connection, is proposed. When using the passive method to control the bit rate of the Internet connection, the minimum amount of traffic is consumed, and the bit rate can be monitored constantly.

To explain the principle of determining the speed in a passive way, consider the principles of operation of the dedicated data link at the physical level (DPDCH). A dedicated data link at the physical layer carries high-level information, including user data, while a dedicated physical layer control channel (DPCCH) carries the necessary control information at the physical layer. The data rate of DPCCH is constant, while the data rate of DPDCH between frames can vary [5, 11]. To separate the signals of one cell from another, scrambling codes and truncated gold sequences are used. All subscriber terminals located in the area of one cell operate at the same frequency, and to separate the transmitted information from the subscriber terminals to the base station and back, Walshcodes are used, with an orthogonality coefficient α [3, 6]. Scrambling codes are superimposed on sewer codes so that they do not change the signal bandwidth, but simply allow signals from different sources to separate from each other. If there is a large delay in the radio channel, the subscriber terminal will perceive part of the transmitter signal of the base station as an obstacle to multiple uses. Therefore, in practice, it is very difficult to obtain an ideal orthogonal signal (typical value). The signal output power of the base station antenna is equal to:

$$P_{\text{TX}} = \frac{N_{\text{rf}} \overline{L} \sum_{j=1}^{N} \upsilon_j \frac{(E_{\text{b}}/N_0)}{W/R_j}}{1 - \overline{\eta_{\text{DL}}}} \tag{5}$$

where W—chip transfer rate, bit/s; R_j—bit rate of the jth subscriber terminal (ST), bit/s; υ_j—the activity coefficient of the jth user at the physical level, for the language—0.67, for data transmission—1; $\overline{L}$—average attenuation of the information signal between the cell transmitter and ST; N—number of ST serviced by a cell; E_{b}/N_0—the energy of the information bit divided by the spectral power density of the noise; N_{rf}—noise power at the input stage of the receiver ST:

$$N_{\text{rf}} = N_{\text{m}} + \text{NF}$$

where N_{m}—thermal noise level. At a bit rate of 3.84 MChip/s; $N_{\text{m}} = 108.2$ dB—noise ratio of the receiver ST. Typical values = 5–9 dBm; $\overline{\eta_{\text{DL}}}$—average load factor in the downlink, which can be found as:

$$\overline{\eta_{\text{DL}}} = \sum_{j=1}^{N} \upsilon_j \frac{(E_{\text{b}}/N_0)_j}{W/R_j} \left[(1 - \overline{\alpha}) + \overline{i} \right] \tag{6}$$

where $\overline{\alpha}$—the average coefficient of the base station transmitter signal orthogonality for the jth ST; $\overline{i}$—the ratio of the level of interference from another cell to interference in its own cell. The ratio E_{b}/N_0 defines the minimum requirements for the radio link to be able to receive/transmit data. For speech transmission $E_{\text{b}}/N_0 = 5$ dB is sufficient, and for data transmission at a speed of 384 kbps the ratio $E_{\text{b}}/N_0 = 1$ dB is necessary [8]. With the help of (5) and (6), a ratio is obtained to estimate the bit rate of the kth terminal R_k:

$$R_k = \frac{W \cdot P_{\text{TX}} - W \cdot P_{\text{TX}} \cdot \sum_{j=2}^{N} \upsilon_j \frac{(E_{\text{b}}/N_0)_j}{G_j} \left[(1 - \overline{\alpha}) + \overline{i} \right] - W \cdot N_{\text{rf}} \cdot \overline{L} \cdot \sum_{j=2}^{N} \upsilon_j \frac{(E_{\text{b}}/N_0)}{G_j}}{\upsilon_k \cdot (E_{\text{b}}/N_0)_k \cdot \left[N_{\text{rf}} \cdot \overline{L} + P_{\text{TX}} \cdot \left((1 - \alpha_k) + \overline{i} \right) \right]} \tag{7}$$

where R_k—the bit rate of the terminal understudy, bit/s; υ_k—the coefficient of user activity at the physical level; α_k—the coefficient of orthogonality of the base station transmitter signal for the at understudy; $(E_b/N_0)_j$—the energy of the information bit divided by the spectral noise power density for the jth AT, times; G_j—gain in signal/noise ratio when processing signals of the jth ST.

In formula (7), the kth subscriber terminal is included in the total number of N terminals serviced by one transmitter. In this case, the kth ST is investigated, and the jth ST is served by the same transmitter, but the kth ST is excluded from their number, that is. k—corresponds to $j = 1$. If the base station is serviced by one and stored ST the ideal orthogonality of codes, the expression (7) is simplified as it eliminates the influence of all ST in addition to researched:

$$R_k = \frac{W \cdot P_{\mathrm{TX}}}{\upsilon_k \cdot (E_b/N_0) \cdot \left[N_{\mathrm{rf}} \cdot \overline{L} + P_{\mathrm{TX}} \cdot \left((1-\alpha_k) + \overline{i}\right)\right]} \tag{8}$$

Let us calculate the value of the bit rate for the typical network parameters. Let the power of the base station P_{TX} is 16 W (42 dBm), the signal power in the receiver is −80 dBm, the average attenuation on the track is (L = 42 + 80)·122 dB, the spectral noise density in the input stage of the receiver of the subscriber terminal $N_{\mathrm{rf}} = -101$ dBm, it is used for data transmission, then $i = 0.55$, which corresponds to the macrocell with non-directional antennas. For such parameter values, the bit rate is equal to:

$$R_k = \frac{3.84 \cdot 10^6 \cdot 16}{1 \cdot 10^{1/10} \cdot \left[10^{-101/10} \cdot 10^{122/10} + 16 \cdot ((1-1) + 0.55)\right]}$$

$$\approx 3.649 \cdot 10^5 \approx 364 \text{ kbps}$$

From the expressions (7) and (8) it can be concluded that the bit rate is directly proportional to the signal power, as well as inversely proportional to the losses on the radio and interference signals arising from violations of the channel codes orthogonality.

$$R_k \approx \frac{W \cdot P_{\mathrm{TX}}}{\left[N_{\mathrm{rf}} \cdot \overline{L} + P_{\mathrm{TX}} \cdot \left((1-\alpha_k) + \overline{i}\right)\right]} \tag{9}$$

Violation of orthogonality occurs not only from the repeated use of frequency, but also due to the multipath propagation of the signal. When the base station serves one ST, it is still possible to break the orthogonality of the codes. For Fig. 3 the dependences of the bit rate R_j on the orthogonality coefficient of the signal α at different power of the received signal α obtained by mathematical modelling are presented.

From Fig. 7, it can be seen that the bit rate depends on the violation of orthogonality of the channel codes. The higher the level of the received signal, the more clearly this dependence is expressed. For example, at the signal level of −75 dBm, there is a change in the bit rate by 20%, when the orthogonality coefficient $\overline{\alpha} = 0.3 \ldots 0.9$ changes. From the ratio (7) it can find the bit rate R_j as:

$$R_j = \frac{(1 - \overline{\eta_{\mathrm{DL}}}) \cdot P_{\mathrm{TX}} \cdot W}{N_{\mathrm{rf}} \cdot \overline{L}_j \cdot \upsilon_j \cdot (E_b/N_0)} \tag{10}$$

From the expression (10) it follows that the bit rate of *j*th AD will tend to 0 when η_{DL} approaching 1 (Fig. 8).

According to the ratio (9), the average load factor in the downlink increases with the number of subscribers, as shown in Fig. 7. It is seen that as the load factor

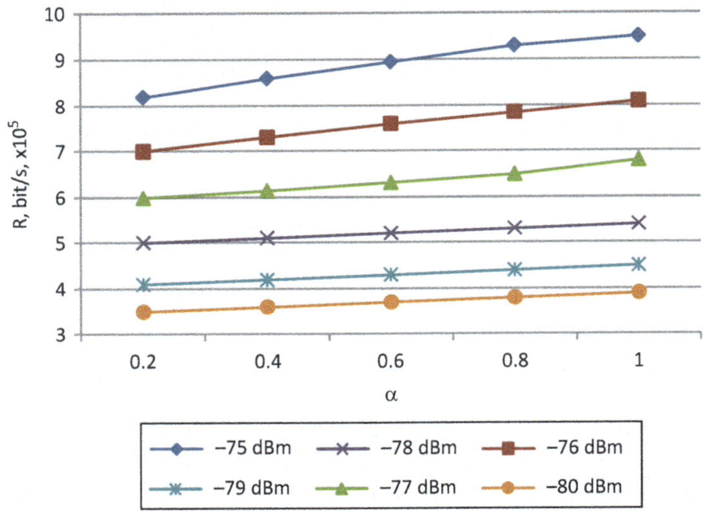

Fig. 7 The dependence of the bit rate on the orthogonality coefficient of the terminal understudy. (Source: [5])

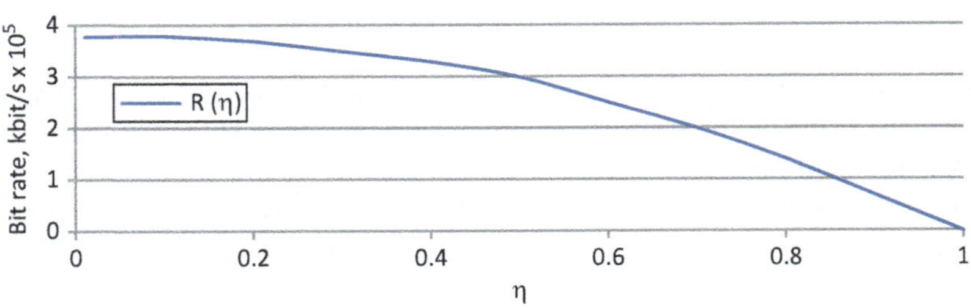

Fig. 8 Dependence of bit rate on load factor. (Source: [5])

increases, the bit rate decreases, orthogonal interference increases. In accordance with the growth of the total number of subscribers, the bit rate of the *j*th ST decreases.

Let assume that the code orthogonality violation through multipath propagation for the studied subscriber is equal, and the code orthogonality violation through multiple frequency use will vary from 0.1 to 0.8. In this case, it can be taken a typical case of using the network, when the terminal understudy is in data transfer mode, and the other ST is used to transmit voice messages. Using the ratio (7) obtained by the author, the authors of the paper will simulate the change in bit rate from the number of subscribers. The results are shown in Fig. 9.

Figure 9 shows that when the number of active subscriber terminals increases, the bit rate of the investigated terminal decreases. This effect is because the load factor increases. At the same time, with the increase in the number of subscribers, the effect of violating orthogonality of codes increases and leads to an increase in the level of interference.

In this regard, the change in the coefficient of orthogonality of sewage ST codes serviced by one transmitter of the base station leads to deterioration of the air condition, and in accordance with the decrease of bit rate. The simulation results are presented in Fig. 10.

The bit rate of the terminal understudy depends on the violation of the terminal codes orthogonality of the understudy caused by the multipath propagation of the signal, as well as the violation of the other subscriber terminals codes orthogonality caused by the numerous use of frequency. Thus violations of subscriber terminals codes orthogonality bring a hindrances additional component in the general air situation of the base station studied sector. Accordingly, if you carry out once in a given period of time monitoring the speed of the file transfer method to the ftp server, set the maximum possible speed for a given location, and then conduct passive speed monitoring with normalization of the results with the maximum value, it can be obtained a daily distribution of the Internet connection speed changes without spending a huge amount of traffic. In addition, this method will not occupy the resources of the radio line and the controller, as well as not to

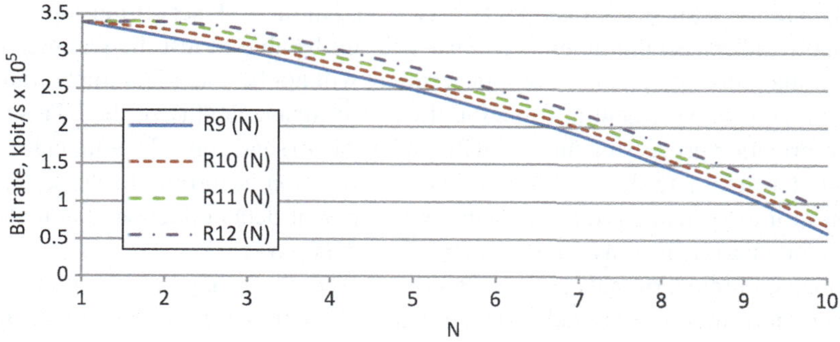

Fig. 9 Bit rate dependence on the number of subscribers. (Source: [8])

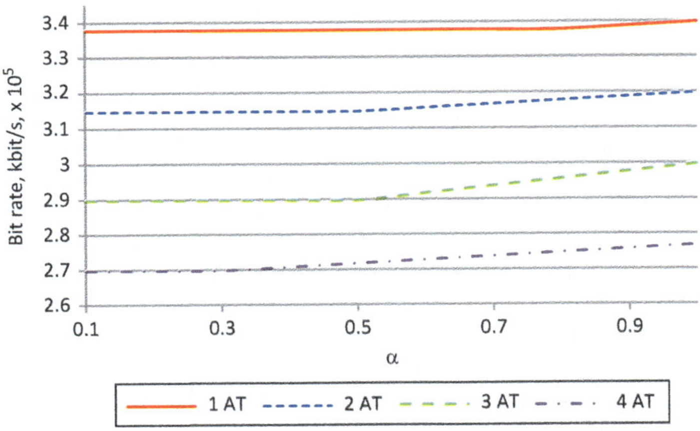

Fig. 10 The dependence of the bit rate of ST on the orthogonality of codes. (Source: [5])

make noise interference in the studied sector, which in long-term measurements will significantly relieve the cellular network.

Let compare the results obtained by active and passive methods of measuring the speed of the Internet connection. In Fig. 11, the diagram shows the results of daily measurements of the bit rate and the calculated values of the bit rate. Figure 11 shows that the maximum load hours are determined in an active and passive way at the same time. The same can be declared about the minimum load hours, they are also determined at the same time. It can be seen that the behaviour of the dependencies is the same, the inflexion points of the decreasing and growing areas are the same in the diagrams obtained in different ways.

It is also worth noting that this method is mainly intended for cellular communication systems continuous monitoring in terms of monitoring the bit rate of the Internet connection.

At the beginning of the system, its launch should begin with the launch of the FTP server, since it stores the general information available for both components of the information and measurement system. Further, when operating the system, the synchronization mode may not be observed, but the FTP server must always be available, otherwise the mobile terminal will not be able to test the parameters of the cellular network. Also, the mobile terminal will not be able to remotely transmit reports with measurement information. In case of unavailability of the FTP server, the information processing module will not be able to send control commands to the mobile terminal and the mobile terminal will function according to the tasks laid down in it at system startup. After the measurement data collection is completed, the mobile terminals transmit the reports to the FTP server.

Since multiple terminals are expected to be used, there may be a situation where two or more mobile terminals will simultaneously transfer encrypted data to an FTP server. A collision may occur during the data transfer process. This problem is solved in several ways. The first of them is at system start it is necessary to

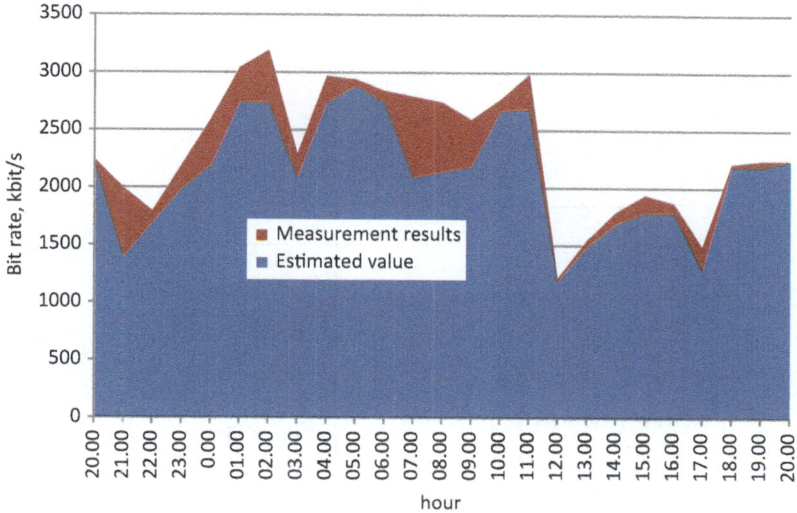

Fig. 11 The calculated and experimental value of bit rate. (Source: Own measurements)

set different time of measurement to all terminals under the condition of their synchronous start. The second is to set different start times for the same time before transferring data to the FTP server. The third is to extend the allowable maximum number of simultaneous FTP server users so that multiple mobile terminals can simultaneously transmit measured data to the server. The best result of solving this problem is the use of a combination of several options, for example, increasing the number of simultaneous FTP server users and setting different data collection times. Thus, even with a very large number of simultaneously operating terminals, the probability that one of the terminals will not be able to quickly transfer the accumulated information is quite small. A block diagram of the algorithm of the system is presented in Fig. 12.

After the data are transferred to the server, the mobile terminals update the configurations and control files. Thus, it is possible to manage each terminal separately, updating configuration files individually for each, and all at the same time, updating one file common to all. The use of the first method is advisable during the commissioning of a small number of mobile terminals or at the same type of study of the entire network. The use of this method is justified when using a large number of active mobile terminals, through which a large number of configuration files appear.

Therefore, the best way to configure and manage mobile terminals is the third, hybrid way. It is a way to manage specialized groups of mobile terminals divided into functionally independent clusters. Clusters are designed in such a way that each of them performs certain highly specialized tasks. For example, two clusters of terminals are created, one of which is configured to measure the parameters of the coverage area of cellular networks of the third and fourth generation in rural areas

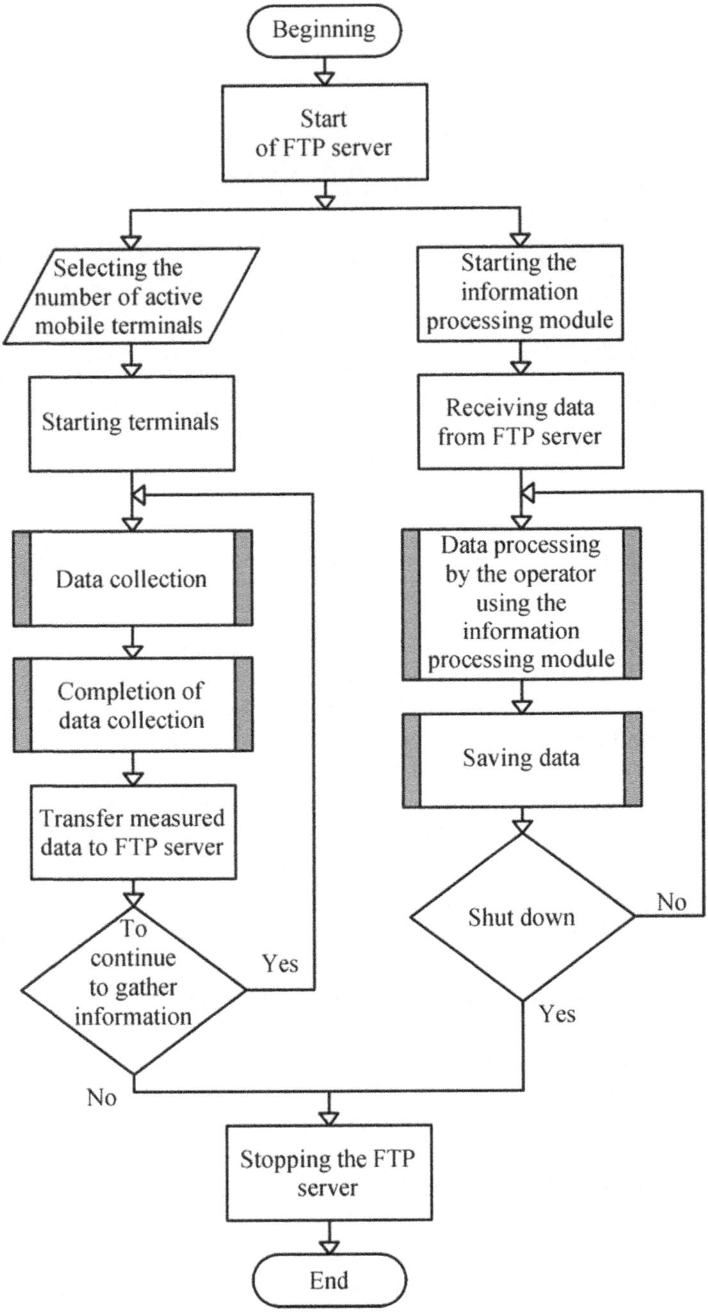

Fig. 12 Block diagram of the algorithm of distributed information and measurement system. (Source: Own measurements)

or on roads of regional importance. The second cluster is configured to measure the bit rate of the Internet connection in the residential and business areas of the city. Settings from the server will be updated separately for the first and second cluster, if necessary, change the number of mobile terminals belonging to one cluster and to another, directly in the process of operation of the system.

The information-processing module is equipped with a graphical interface. It runs an operating system. Configuration files for working with the module are stored locally on the same personal computer where the module is installed. However, it is possible to save configuration files to an FTP server. The latter allows you to use the information-processing module with the same settings on different personal computers.

This is quite convenient when you need to process volumes of information in one place, such as the workplace of the operator conducting the measurements, and provide the results to another place, such as the organization for the report to the lead specialist. In this case, you can process all the measured material in the workplace, then make a report, and save the configuration files on a remote server. The information-processing module is capable of simultaneously processing information from all active mobile terminals.

To deliver all measured data to the local file storage, it is enough to synchronize the data with an FTP server. Synchronization is performed by checking the existing files in the storage and on the local computer. Files that are missing from the local computer will be copied from the remote FTP server. Thus, the data prepared for processing will be automatically transferred to the local computer for processing. A group of files is processed in the same way as a single file.

When the system is shut down, the FTP server should be the last to shut down, because without it the system will function only in fragments.

5 Conclusions

The main objective of the bit rate study in a passive way is to identify trends in user behaviour, namely the hour's identification of maximum and minimum load on the base station.

When carrying out daily measurements in cellular networks, the dynamics of the studied process is more important than the accuracy of measurements. It is important to identify the moments when the bit rate starts to decrease, or to grow, it is important to make the correction in the operation of the network, thus maintaining the quality of services at a high level. Based on data on dynamics of bit rate change, it could be received the ability to reallocate dynamically network resources. For example, during the day in the hours of minimum load can reduce the power of the transmitter, thereby saving electricity. When using the dynamic mode at one base station, the effect of energy-saving is unobtrusive, but when using the dynamic mode of the network as a whole gives a significant economic effect. For example, the power consumption of the base station is 500 W h. Let the total power consumption of the

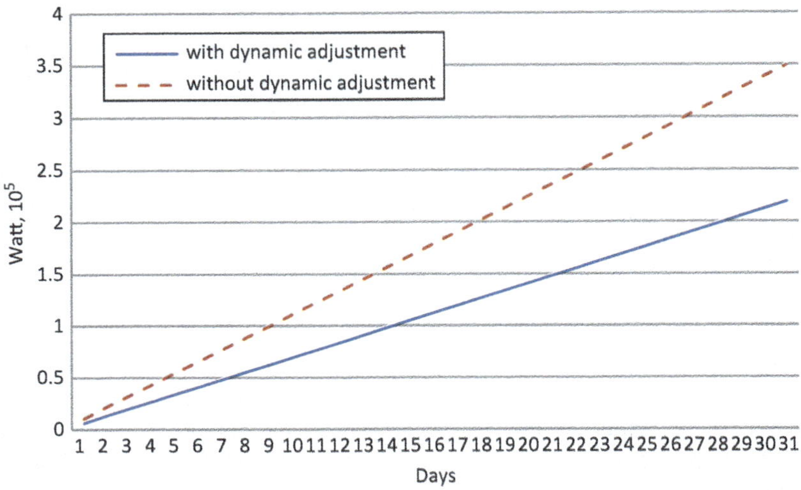

Fig. 13 Total power consumption of the base station. (Source: Own measurements)

control and cooling systems be 300 W h when the transmitter power is reduced by two times. Calculate the total amount of electricity consumed per month by day, the simulation results will be presented in the form of graphs in Fig. 13.

Figure 13 shows that the total energy consumption of one base station without dynamic power control is 0.35 MW per day, and with dynamic control is 0.2 MW per day. Energy savings from one base station can be up to 150 kW per day. One mobile operator in the region has about 150–200 base stations. Thus, the total energy savings range from 22.5 to 30 MW per month. Thus, the use of dynamic power control can reduce energy consumption, thereby reducing heat generation and improving the operating conditions of base stations of mobile operators. This technique has been tested on 2G and 3G networks, but it is suitable for use in modern cellular networks such as 4G and 5G.

Acknowledgements This paper has been written within VEGA project Nr. 2/0077/19: Work competencies in the context of Industry 4.0 development and KEGA project Nr. 030STU-4/2018: E-platform for improving collaboration among universities and industrial enterprises in the area of education.

References

1. MobUA Homepage, https://www.mobua.net/maps/?pos=48,31,6. Last accessed 18 Dec 2019
2. D. Cagáňová, M. Bawa, J. Šujanová, A. Saniuk, *Innovation in Industrial Enterprises and Intercultural Management*, 1st edn. (University of Zielona Góra, Zielona Góra, 2015), 126 p. ISBN: 978-83-933843-4-1

3. P. Bakhovsky, M. Yevsyuk, Some aspects of mobile telecommunication development. Adv. Technol. Dev. **13**, 25–32 (2018)
4. Y. Toroshanko, Y. Selepyna, N. Yakymchuk, V. Cherevyk, Control of traffic streams with the multi-rate token bucket, in *International Conference on Advanced Information and Communications Technologies, AICT 2019*, 2019, pp. 352–355. https://doi.org/10.1109/AIACT.2019.8847860
5. H. Holma, *W-CDMA for UMTS* (John Wiley & Sons Ltd., Chichester, 2004)
6. M. Naworocki, M. Dohler, A. Aghvami, *Understanding UMTS Radio Network. Modelling, Planning and Automated Optimization* (John Wiley & Sons Ltd., Hoboken, NJ, 2006)
7. N. Yakymchuk, P. Stanko, Correlation-regression analysis of information networks with non-stationary traffic, in *Conference "Actual Problems of Information Technologies: ARIT-2018"*, Kyiv, 2018, pp. 128–130
8. R. Röhrle, I. Ehriel, I. Danilov, GPRS protocols and testing. Commun. Networks Syst. **12**, 44–50 (2002)
9. M. Bawa, D. Cagáňová, Selecting network protocols for internet of things based upon innovation and knowledge management. J. Telecommun. Syst. Manag. **7**(2), 1–4 (2018). https://doi.org/10.4172/2167-0919.1000169. ISSN: 2167-0919. V databáze
10. V. Saiko, T. Naritnik, P. Bakhovsky, Ultra-fast terahertz radio access channel for 4th and 5th generation mobile networks. Technical news. Sci. J. **1**(47), **2**(48), 48–50 (2018)
11. N. Yakimchuk, V. Olianskii, Characteristic of the system of clock network synchronization to estimate the delay time of the network, in *Conference "Problems of Telecommunications"*, Kyiv, 2018, pp. 107–110

Impact of ICT on Emergency Response Vehicle Driving Time in a Vicinity of Large Events

Tibor Petrov, Milan Dado, and Tatiana Kováčiková

1 Introduction

Connected and autonomous vehicles are supposed to bring major benefits in terms of increased traffic safety, efficiency, and decreased environmental impact. The expected benefits are mainly brought by the ability of vehicles to exchange information from their sensors and to communicate their state to the traffic management entities which can take educated decisions to control the traffic flows.

Means of communication between vehicles, infrastructure, and traffic management entities may vary from legacy traffic-light based systems to state-of-the-art electronic communication systems. In any case, it is extremely important to provide a backward compatibility for older vehicles which are not equipped with electronic communication modules. These vehicles are still an important, and for several upcoming years probably major, part of the traffic.

There were several algorithms presented, which use the data from connected vehicles' sensors to allow more efficient intersection management, reducing waiting times of involved vehicles, e.g [1, 2]. According to simulation results, these algorithms can help to minimize waiting time at the intersection and partly address traffic congestion.

T. Petrov (✉) · T. Kováčiková
Department of International Research Projects – ERAdiate+, University of Zilina, Zilina, Slovakia
e-mail: tibor.petrov@uniza.sk; tatiana.kovacikova@uniza.sk

M. Dado
Department of Multimedia and Information-Communication Technologies, Faculty of Electrical Engineering, University of Zilina, Zilina, Slovakia
e-mail: milan.dado@uniza.sk

D. Cagáňová et al. (eds.), *Advances in Industrial Internet of Things, Engineering and Management*, EAI/Springer Innovations in Communication and Computing,
https://doi.org/10.1007/978-3-030-69705-1_2

To tackle the increasing emergency response vehicle (ERV) driving time in congested cities, other relevant systems were either explored or already deployed. Among them, emergency vehicle preemption systems aim at prioritizing vehicles approaching from a certain direction, usually the one of the ERV [3, 4].

Another approach to potentially decrease the ERV's driving time by technological means is by early warning about the ERV's approach [5].

The system, whose feasibility is studied combines both approaches. Early warning for the connected vehicles which can benefit from the rich traffic information received by V2V and I2V communication is studied in conjunction with the principle of preemption by prioritizing traffic flow containing the ERV.

In this chapter, we would like to further explore to what extent Information and Communication Technologies employing Vehicle-to-Everything (V2X) communication and strategically placed Variable Message Signs (VMSs) could help to reduce the driving time of an ERV in a dense urban traffic scenario and what are the requirements for communication, to enable non-delayed transit of such a vehicle. Results from this research could be used as an input for dynamic intersection control algorithms optimized to ensure intersection preemption in a scenario with approaching ERV. In this study, we consider an urban road network during a peak hour near a site where a large cultural or sports event takes place. Driving time of the ERV is evaluated using a realistic simulation of both traffic scenarios as well as communication technologies deployment. In the simulation scenario, we consider a part of a traffic infrastructure near sports stadium in the city of Zilina in Slovakia. Simulations are based on the real traffic infrastructure and vehicle routes.

1.1 Overview of Traffic Control Approaches

According to the European Environment Agency, the passenger transport demand in the European Union increased by 30% between 1995 and 2017 [6]. Passenger cars remain the dominant transport mode with more than 70% share of all passenger transport. To fully utilize the existing transport infrastructure and meet the ever-increasing transport demands of the modern society, efficient traffic control management strategies have to be implemented. Figure 1 depicts the basic elements of a traffic control loop. The behavior of the traffic flow depends on Control inputs and Disturbances. Control inputs are fed to the control actuators (e.g., traffic lights, variable message signs, etc.) and can be manipulated by the traffic management entity. The disturbances (e.g., traffic demand, various traffic incidents, etc.) can be measured, detected, or predicted but cannot be manipulated. The performance of the traffic network is measured by a suitable performance metric. Surveillance helps to extend the measured data captured by sensors (e.g., loop detectors, cameras, even connected vehicles) as an input to the control strategy and provides an assistance to the human operator of the traffic control system. The control strategy lies in the core of the control loop and it is responsible for the real-time definition of control inputs in order to fulfill the pre-specified control goals despite the influence of the

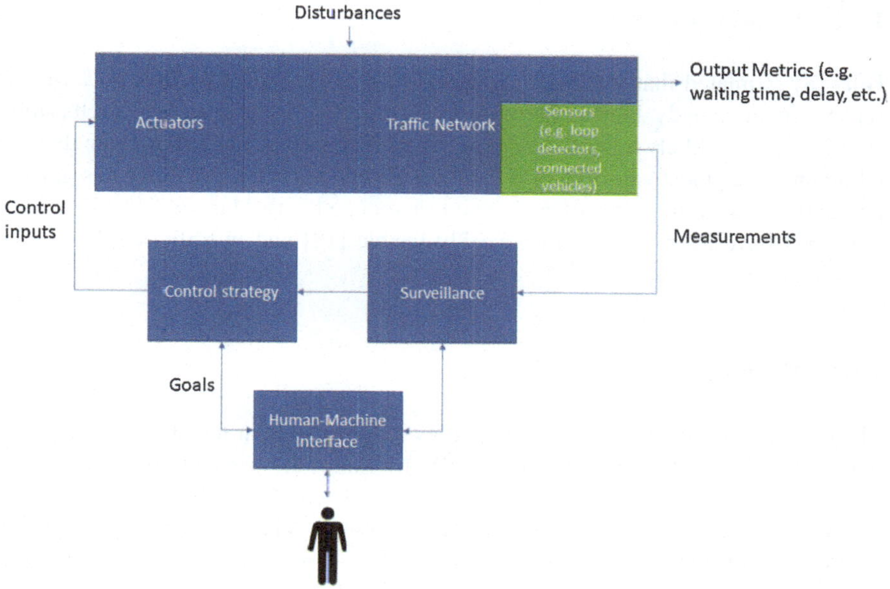

Fig. 1 Basic elements of a traffic control loop [7]

disturbances in the traffic network. The decision process of the control strategy is supported by available measurements, estimations, or predictions.

Intersection traffic control strategies can be divided according to various criteria. Considering the action scale of the employed traffic control mechanism, isolated or coordinated traffic control strategies are widely used.

Isolated strategies focus on controlling a single intersection, while coordinated strategies are used to control a wider urban area or even a whole network consisting of many intersections.

In the context of adaptability of a control strategy to the current traffic flow, fixed-time or traffic-responsive control strategies can be employed. Fixed-time control strategies often use historical data to derive intersection control parameters for a given time of day. The timing of the control signals remains constant for an extended time period. Traffic-responsive strategies use real-time measurements of the traffic situation (e.g., by inductive-loop detectors) to optimize the traffic control parameters. At the expense of necessity of an additional measurement infrastructure, they can achieve higher efficiency in satisfying the instantaneous traffic demand [7].

1.2 *Overview of Current Emergency Vehicle Warning Systems*

Current emergency vehicle warning systems can be divided into three basic groups. These are almost always combined to make strong visual and audible impact on the driver's senses.

1.2.1 Visual Warning Systems

Hills [8] estimated that a driver in a motor vehicle obtains more than 90% of the sensory input visually. Furthermore, several studies [9, 10] have shown that flashing signals are more likely to attract a driver's attention than steady lights. Even though a human eye is most sensitive in the yellow-green spectrum, red color has gained popularity as a warning identifier since it is associated with "danger" [11]. However red lights have been shown to be weakly visible [12] and in traffic easily lost in vehicle tail lamps [13].

1.2.2 Markings

The markings of an emergency vehicle play a crucial role in the ability of a driver to detect and identify the approaching ERV. Markings are particularly important in a case of ambulances which resembles commercial vehicles [14].

However, Allen [12] argues that a single-colored vehicle is more likely to be visible than a vehicle with a multitone body color. This effect is even more obvious in an urban environment according to findings by Hills [8, 15].

1.2.3 Audible Warning Systems

Sirens have been used for a long time as an emergency vehicle warning system. Recommended characteristics of an audible warning signal include sufficient power and a wide frequency spectrum for overcoming a masking noise [16], fast variation of pitch [13], and relatively rapid cycling time [17]. The recommended frequency range is between 1 and 4 kHz [18].

Siren, in order to be effective as a warning device, has to compete with the noise generated by the traffic flows, ambient noises, and other sources of audible noise. Some studies [17, 19] claim that an effective range of a typical siren used as emergency vehicle warning system yields from 8 to 12 m, when considering car's noise insulation characteristics.

Studies demonstrate that an efficiency of a siren as a warning device is severely limited. It is effective only at low speeds and on a very short range. Moreover, a siren mode (e.g., wail, yelp, high-low) does not significantly affect the efficiency of the siren.

1.3 Overview of Cooperative ITS and ITS Applications

The ITS architecture consists of three main communication domains – the in-vehicle, V2X domain, and the infrastructure domain. The high-level ITS architecture can be seen in Fig. 2 [20].

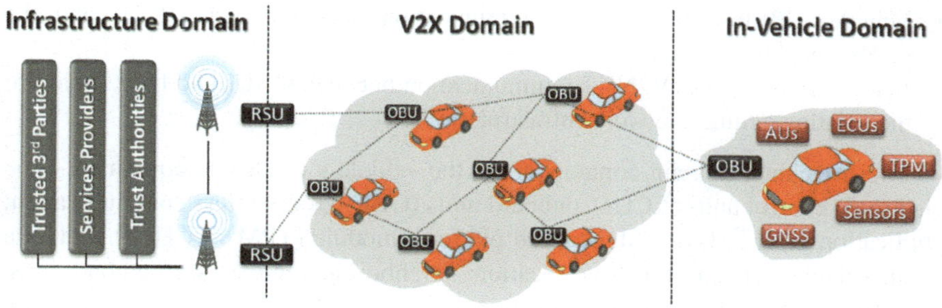

Fig. 2 The high-level ITS architecture [20]

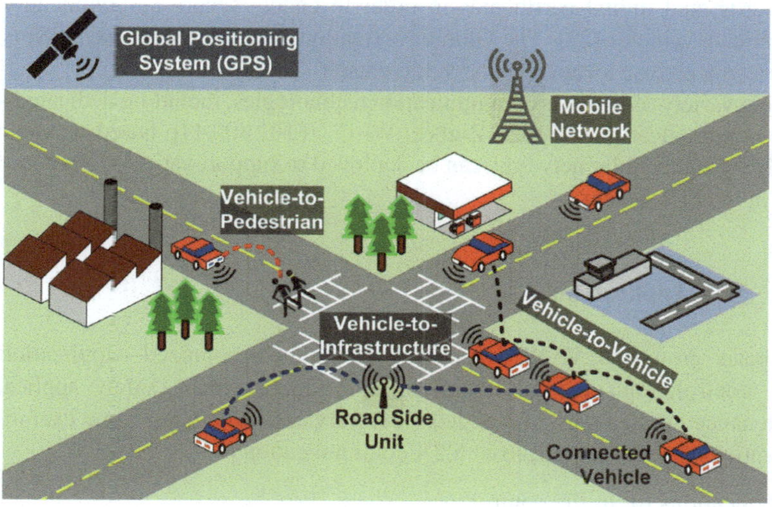

Fig. 3 V2X communication in ITS environment [20]

The infrastructure domain consists of the trusted third parties (e.g., vehicle manufacturers), service providers, and the trust authorities. Trust authorities provide registration and authentication of roadside units (RSUs) and onboard units (OBUs). The service providers provide applications and connectivity to the vehicle's application units (AUs) as well as the software updates, billing, and other services. RSUs interconnect the V2X and the infrastructure domain.

Vehicle OBUs and RSUs form the V2X domain, which is often called the ad hoc domain. Data from previous communication and vehicle sensors can be stored in the OBU and then transmitted to other OBUs, RSUs, or pedestrian communication units by wireless communication means as can be seen in Fig. 3 [20]. Four main types of vehicular communication technologies can be distinguished:

- Vehicle-to-Vehicle (V2V)—direct communication between nearby OBUs
- Vehicle-to-Infrastructure (V2I)—communication between OBUs and RSUs

- Vehicle-to-Pedestrian (V2P)—communication between OBUs/RSUs and the pedestrians in the range
- Vehicle-to-Network (V2N)—communication between OBUs and base station of the existing cellular network infrastructure

The in-vehicle domain contains all of the vehicle's electronic equipment—the electronic control units (ECUs), onboard units (OBUs) for wireless communication, application units (AUs), and a trusted platform module (TPM). ECUs collect data about vehicle's speed, direction, location, neighboring vehicles, traffic conditions, etc., and control the vehicle's functionality [20]. AUs are capable of running applications provided by the service providers using the OBU communication module [21]. The TPM is a secure cryptoprocessor which uses cryptographic keys, passwords, and digital certificates in order to enable secure communication and authenticate vehicles [22]. The Global Navigation Satellite System (GNSS) is used to retrieve a precise location of the vehicle itself.

A big variety of wireless communication technologies, including dedicated short-range communication (DSRC), Zigbee, Wi-Fi, IEEE 802.11p-based technologies, Wi-MAX, and cellular networks can be deployed to support various communication tasks in the ITS system.

1.3.1 V2X Applications

Five basic groups of VANET applications can be recognized—applications for traffic control, logistics and freight transport management, safety applications, maintenance applications, and added value services. Below are some examples of applications that can benefit from V2X communication [23].

1. Applications for traffic control
   - Route planning
   - Dynamic traffic signs displaying
   - Congestion-based route charging
   - Prioritizing various modes of transport
2. Logistics and freight transport management
   - Parking zones management
   - Route control
   - Dangerous goods transport control
3. Safety applications
   - Cooperative maneuvering
   - Emergency transmission
   - Danger warning
   - Coordinated braking

4. Maintenance applications
   - Sensor calibration
   - Remote diagnostics
5. Added value services
   - eCall emergency call system
   - Advanced navigation
   - Insurance services
   - Stolen vehicle tracking

Raya and Hubaux in [24] categorize applications for VANETs into two categories

- Safety-related applications which are related to the situations, where application presence may prevent life-endangering accident
- Other applications without direct impact on vehicle and crew safety

Cheng et al. in [25] divide applications into three categories:

- Infotainment delivery which offers convenience and comfort to the crew
- Road safety to prevent accidents
- Traffic monitoring and management to maximize road capacity and avoid congestions

For safety-related applications, various grades of performance demands can be required. Table 1 quantifies these parameters [26].

Table 1 Performance parameters for safety-related applications [26]

Performance class	Latency (ms)	Packet generation frequency (p/s)	Communication range (m)	Application
Low latency, high frequency	≤100	10–20	≤150	Accident, control loss, cooperative collision warning
Medium latency, medium frequency	≤200	5–10	≤100–130	Intersection collision warning, lane overtake assistance, extended brake signaling
High latency, low frequency	≤1000	1–2	≤1000	Work zone warning, road condition warning

1.4 Communication Technologies for Connected Vehicles

To allow vehicular data communications, several communication technologies emerged which are described in this section. Two major families of technologies are dominant:

- Dedicated Short-Range Communications (DSRC)
- Cellular Vehicle-to-Everything (C-V2X)

Communication technologies based on the DSRC use the IEEE 802.11p communication standard to enable device-to-device communication [27]. Communication nodes (e.g., vehicles and roadside infrastructure) form a Vehicular Ad-hoc Network (VANET). VANET is a self-formed network without centralized management. Communication nodes exchange data using a predefined communication channel whenever they are in the communication range. There are two types of communication channels defined for VANETs—a Control Channel (CCH) for safety-related communications and one or more Service Channels (SCHs) for non-safety-related services. Channels use the 5.9 GHz frequency band.

C-V2X technologies emerged as an extension of the LTE standard in 2017 [28]. These technologies use existing cellular network infrastructure to allow vehicular communications. To address V2V and V2I communication scenarios where vehicles are outside of the base station's coverage a new radio interface (PC5) was developed. Since the vehicles do not use downlink nor uplink to exchange data the new communication mode, referred to as sidelink, was introduced.

The network resources are usually allocated by the base station. When the base station is not able to assign resources to vehicles (i.e., in a case of connection loss), the vehicles select communication network resources in the form of resource blocks autonomously [29].

As the C-V2X family of technologies was introduced relatively recently, it has to undergo extensive field testing to prove its operational readiness. For this reason, we decided to study the impact of DSRC-based technologies which are more mature technology [30].

From all candidate technologies, two families of DSRC-based C-ITS standards emerged as dominant:

- IEEE WAVE
- ETSI ITS

Both families of standards use the same IEEE 802.11p physical (PHY) and medium access (MAC) architecture layers (see Fig. 4). Lower communication layers are based on WiFi standard with channel bandwidth reduced to 10 MHz and channel-switching guard intervals doubled.

The key difference between the two families of standards is in their approach to addressing the communication channel congestion. IEEE WAVE uses multichannel operation, i.e., the transceiver is constantly switching between CCH and SCH every 50 ms. The main disadvantage of this approach is rather an inefficient channel

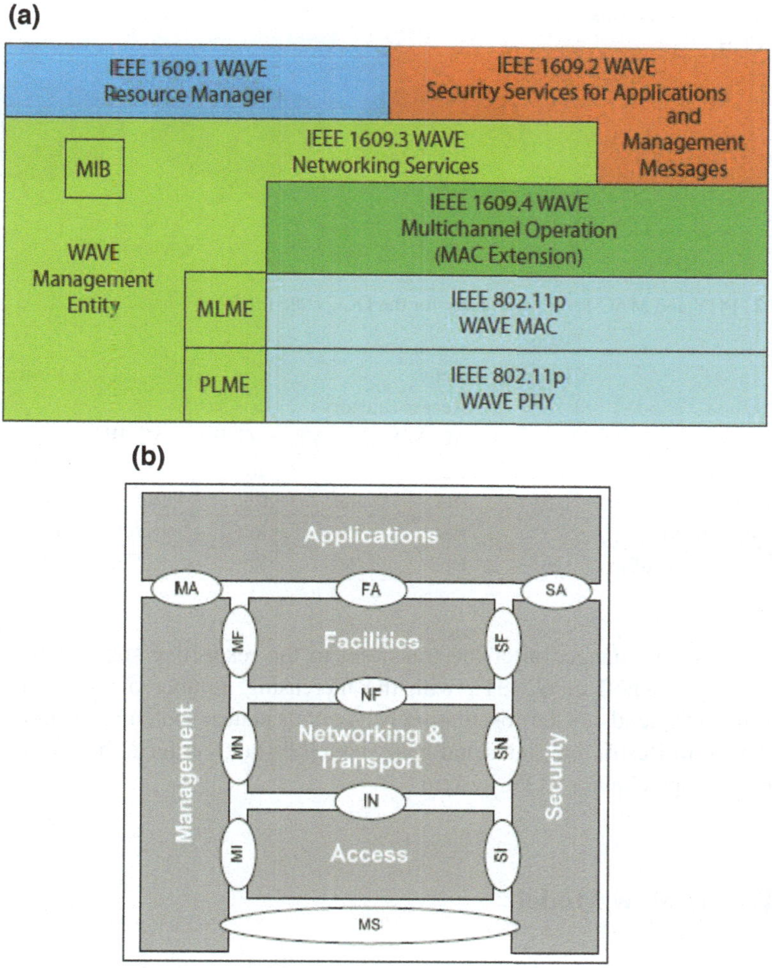

Fig. 4 Architectures of major DSRC-based C-ITS communication technologies. (**a**) IEEE WAVE; (**b**) ETSI ITS

utilization. While multichannel operation allows the use of a single transceiver for both safety-related and other communications, the peak channel utilization is only around 46% considering equal use of CCH and SCH and switching guard intervals. ETSI ITS uses Decentralized Congestion Control (DCC) mechanism to reduce the CCH load in the case of channel congestion. DCC is a state machine which switches between its states according to the detected busy time of the medium. Figure 5 shows the states of the DCC for CCH with 15% and 40% busy time being thresholds when a state change takes place [31]. Each state of the state machine has defined PHY and MAC parameters which are then applied to the system to reduce the CCH load. These parameters are shown in Table 2. The value *ref* means that the corresponding

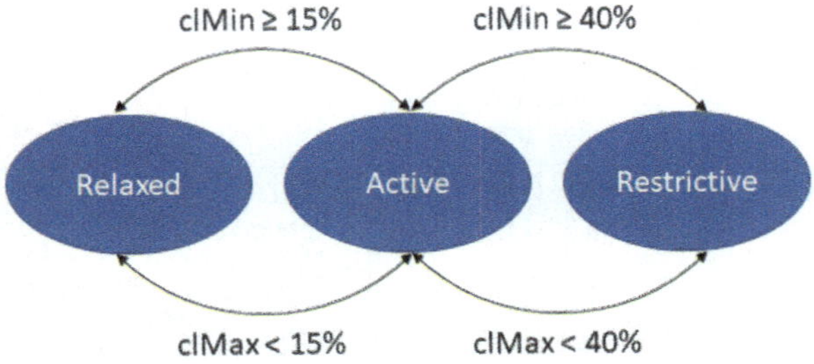

Fig. 5 DCC state machine for the CCH

Table 2 PHY and MAC layer parameters for the DCC state machine

	State					
	Relaxed	Active				Restrictive
		Access category				
		AC_VO	AC_VI	AC_BE	AC_BK	
Power (dBm)	33	Ref	25	20	15	−10
Interval (s)	0.04	Ref	Ref	Ref	Ref	1
Data rate (Mbps)	3	Ref	Ref	Ref	Ref	12
Radio sensitivity (dBm)	−95	Ref	Ref	Ref	Ref	−65

parameter is not changed after the transition to the respective state. However, as observed by Eckhoff et al. [31], with the increasing number of communicating nodes the DCC tends to start oscillating between its states, switching from Relaxed over Active to Restrictive states and back, which instantly affects the channel load forming a loop behavior of the system.

2 Traffic Flow Model

To model the traffic flows, the Simulation of Urban Mobility (SUMO) [32] was used. A real map of the road infrastructure from openstreetmap.org [33] was exported and traffic flow densities were modeled according to the Poisson distribution:

$$p(x) = \frac{\mu^x e^{-\mu}}{x!} \tag{1}$$

where $p(x)$ is the probability for x vehicles arriving in a time interval, and μ is the expected rate of vehicle arrival in that interval.

Traffic scenario was selected in a way to represent the worst possible situation, when the road network is severely congested. We assume a scenario, when a cultural or sport event ends at a stadium and a lot of vehicles try to depart from the parking lot in a limited time. In this situation, an emergency vehicle has to pass the road segments congested by vehicles leaving the stadium's parking lot on its way to the nearest hospital.

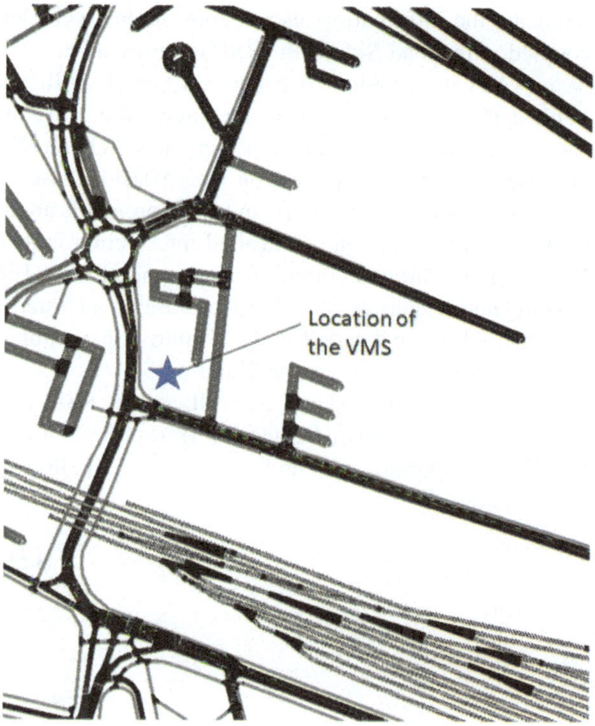

Fig. 6 Part of the simulated road infrastructure

Where possible, traffic densities in road segments surrounding the stadium were modeled according to mean values from a traffic survey conducted for the Traffic General Plan of the City of Zilina [34]. The resulting traffic flows in the simulation consist of vehicles generated according to the survey mean values and the vehicles departing from the stadium's parking lot.

Part of the simulated road network can be seen in Fig. 6. The VMS with RSU capability is located at the intersection of main road which an emergency vehicle has to pass and a side road leading from the stadium's parking lot. There is also a flow of vehicles approaching from the city ring through a roundabout at the main road. These vehicles do not have a visual contact with the VMS, but they are in the communication range of the VMS's RSU. If the vehicle entering the roundabout is connected, it can receive instructions via an electronic communication module, preventing the congestion of the roundabout.

3 Communication Architecture

Since it is expected that during upcoming years the traffic flow would consist of a mix of both connected and conventional vehicles without electronic communication capacity, the simulation experiment uses Variable Message Sign (VMS) to communicate the information also to conventional vehicles. In addition, used VMS is coupled with Road Side Unit (RSU) to provide more detailed control data beyond basic VMS information to connected vehicles. To disseminate Infrastructure-to-Vehicle (I2V) communication messages, the RSU uses 802.11p-based protocol stack (Fig. 7) running scalable Emergency Vehicle Warning Application (EVWA) proposed in [5]. The application uses 802.11p-based physical layer and medium access layer. On top of that, IP and UDP protocols are used for node addressing and packet routing. Since the controlled intersection is located in close vicinity of the RSU, only the limited communication range for the EWVA is required. Therefore, in simulated scenarios, EVWA packets are broadcasted and multi-hop transmissions are not used. More detailed study of multi-hop communication delays in the context of the proposed EVWA can be found in [5].

The VMS display panel as well as the RSU are controlled remotely from the Traffic Management Center via Transport Control Protocol (TCP) connection. Simulated communication architecture with data flow can be seen in Fig. 8.

Fig. 7 RSU and vehicles' protocol stack used for simulation

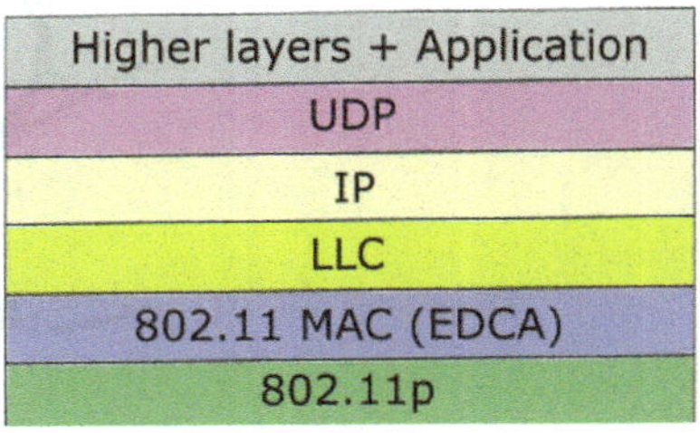

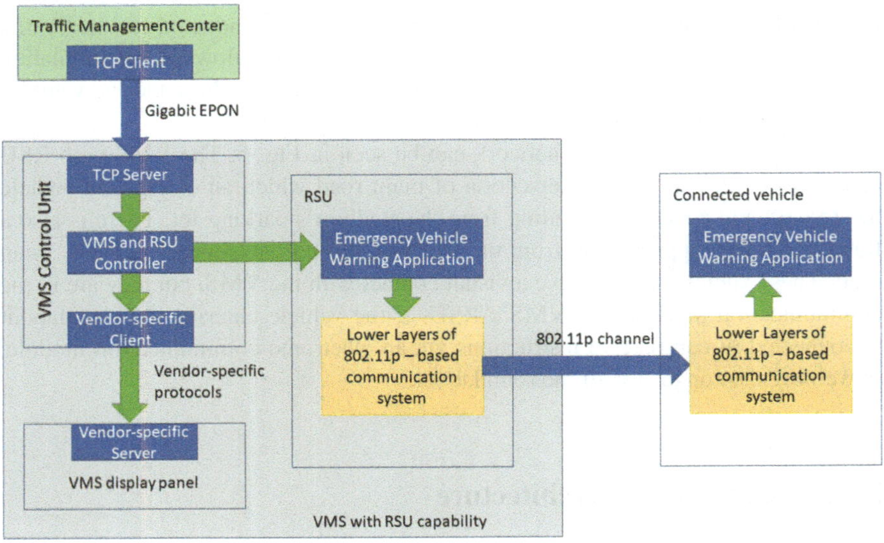

Fig. 8 Communication architecture of the simulated traffic control system

In the simulation of the communication network, VMS vendor-specific protocols were considered to be free from delay and any data loss.

The properties of the communication network were simulated using OMNeT++ discrete event network simulator [35]. Communication delays acted as an input to the traffic simulator for calculation of the emergency vehicle's driving time.

4 Simulation Scenarios

For the purpose of the simulation we considered three types of vehicles:

- Conventional vehicle—vehicle with human driver without equipment to receive any electronic communication
- Connected vehicle—vehicle with human driver equipped with communication module fully compatible with RSU
- Emergency response vehicle—ambulance with human driver equipped with communication module capable of message exchange with the Traffic Management Center

Using the abovementioned three types of vehicles, four simulation scenarios were considered:

- No communication between vehicles and Traffic Management Center (TMC)
- Specifically designed Variable Message Signs (VMS)
- Isolated RSU without any means to communicate with conventional vehicles
- Infrastructure-To-Vehicle (I2V) communication enabled VMS

The impact of the communication on emergency vehicle's driving time was evaluated for varying penetration of connected vehicles from 0% to 100%. The emergency vehicle's route length was 3600 m. Part of the route with a length of 2100 m is covered by an uncongested city ring. The length of the emergency vehicle's route through the city streets is 1500 m—this is also the part of the journey, where undesirable delay in driving time is gathered.

The Emergency Vehicle Warning application simulated in the scenarios does not only provide the information about approaching emergency vehicles, but also transfers commands with a required action to the connected vehicles (beyond Day 1 C-ITS application [36]). In all simulation experiments, we assume drivers fully obey required actions.

5 Simulation Results

The driving time of the emergency vehicle without any intersection management was 217.7 s. As a next step, we studied the emergency vehicle's delay compared to free-flow conditions at the intersection. In this scenario, we assumed 100% of the

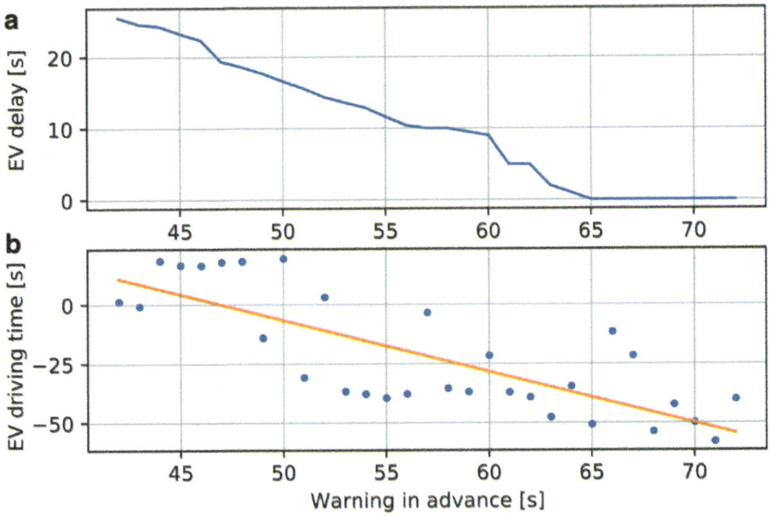

Fig. 9 Dependency of reduction of emergency vehicle's delay (**a**) and driving time (**b**) on advance warning time

vehicles, which have visual contact to the VMS will obey the traffic management request presented at the VMS. For this purpose, we treated both the connected and conventional vehicles the same way. As illustrated in Fig. 9a, in the studied scenario the delay which emergency vehicle experiences is dependent on how much in advance the information about approaching emergency vehicle is offered to regular vehicles. If drivers received the information more than 65 s in advance the emergency vehicle did not experience any delay. Total driving time of the emergency vehicle is reduced according to Fig. 9b. Blue dots represent the values from individual experiments, while the trend line is depicted in orange.

The impact on driving time lies in the range from −25% to +10%. The positive numbers mean, the system can even introduce a loss under specific conditions. This is due to fluctuations in the traffic flow that occurs when certain vehicles, due to their earlier arrival thanks to the preemption system, have to wait at the previous intersection for the green light and form an obstacle for the emergency vehicle. This issue indicates that traffic management in favor of emergency response vehicles should be done in a coordinated manner on a number of consecutive intersections along the route of the emergency vehicle rather than for isolated intersections.

When only isolated RSU is used for message exchange, the percentage reduction of emergency vehicle's driving time, dependent on the penetration rate of equipped vehicles, can be seen in Fig. 10. Note, that in this scenario, only vehicles with V2X communication capability were able to receive the traffic management information. The connected vehicles were selected randomly according to the uniform distribution. In the studied scenario, we assumed the information about the approaching emergency response vehicle is available 30s before its arrival. According to the

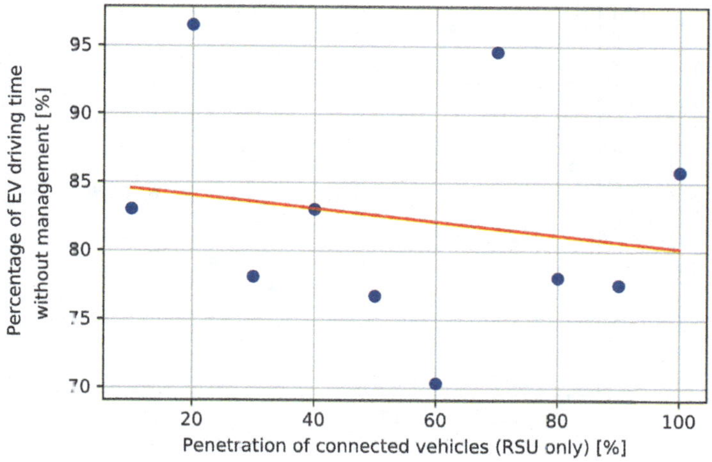

Fig. 10 Relation between the reduction of emergency vehicle's driving time and penetration of connected vehicles in the scenario with no means of communication between TMC and conventional vehicles

survey made among professional emergency response vehicles drivers in Germany [37], this is the appropriate reaction time needed for other drivers to correctly prepare for the emergency vehicle's approach.

The reduction of emergency vehicle's driving time can be expected in the range from 3% up to 30% for some specific cases.

The same traffic scenario was simulated also with the employment of VMS with integrated RSU. In this simulation scenario, all vehicles in visual contact with the VMS were considered to be able to comply with the requested traffic management operations, while vehicles on nearby roads were informed only via electronic communication. The dependence of emergency vehicle's driving time reduction for different penetration of connected vehicles on nearby road segments can be seen in Fig. 11. The expected reduction of the emergency vehicle's driving time ranges from 6% to more than 25%.

6 Conclusion and Further Work

In the chapter, a scenario of approaching emergency vehicles under specific conditions of congested road network in a vicinity of a large cultural or sports event was simulated. The impact of the deployment of Information and Communication Technologies (ICT) on the emergency vehicle's driving time was studied.

The results of the simulation experiments confirm that ICT deployment can reduce the emergency vehicle's driving time in dense urban traffic scenarios; however, the results also reveal, that if deployed in an isolated manner, the system can

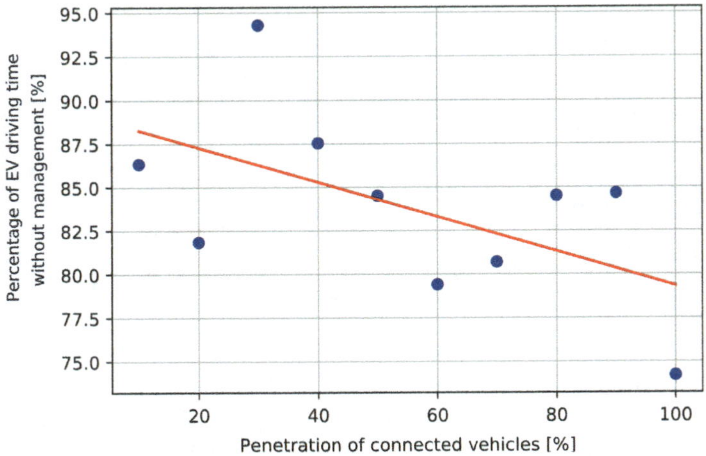

Fig. 11 Relation between the reduction of emergency vehicle's driving time and penetration of connected vehicles in the scenario with VMS and RSU communication employed

even introduce a loss. To foster the benefits of the intersection preemption system for support of emergency response vehicles, several consecutive intersections have to be controlled in a synchronized and optimized way.

In our future work, based on previous research in the field of vehicular networking [38–40], we would like to design communication architecture to support innovative intersection control algorithms for minimizing the driving time of emergency vehicles.

Acknowledgments Research described in the chapter was facilitated by the ERAChair ERAdiate—Enhancing Research and innovation dimension of the University of Zilina in intelligent transport systems, Grant agreement no: 621386, Project 7thFP.

This work was supported by the Slovak Research and Development Agency under the contract no. SK-IL-RD-18-005.

References

1. S.I. Guler, M. Menendez, L. Meier, Using connected vehicle technology to improve the efficiency of intersections. Transp. Res. C Emerg. Technol. **46**, 121–131 (2014). https://doi.org/10.1016/j.trc.2014.05.008
2. W. Liu, G. Qin, Y. He, F. Jiang, Distributed cooperative reinforcement learning-based traffic signal control that integrates V2X networks' dynamic clustering. IEEE Trans. Veh. Technol. **66**, 8667–8681 (2017). https://doi.org/10.1109/TVT.2017.2702388
3. Y. Wang, Z. Wu, X. Yang, L. Huang, Design and implementation of an emergency vehicle signal preemption system based on cooperative vehicle-infrastructure technology. Adv. Mech. Eng. **2013** (2013). https://doi.org/10.1155/2013/834976
4. M. Miyawaki, Z. Yamashiro, T. Yoshida, Fast emergency pre-emption systems (FAST), in *Proceedings 199 IEEE/IEEJ/JSAI International Conference on Intelligent Transportation Systems*, (IEEE, Piscataway, NJ, 1999), pp. 993–997

5. T. Petrov, *Cooperative Technologies in Intelligent Transportation Systems* (University of Zilina, Zilina, 2018)
6. European Environment Agency. Passenger and freight transport demand in Europe, https://www.eea.europa.eu/ds_resolveuid/653cb59bd80f49de891d49195f159971/. Last accessed 20 Jan 2020
7. M. Papageorgiou, Overview of road traffic control strategies. IFAC Proc. Vol. **37**(19), 29–40 (2004)
8. B.L. Hills, Vision, visibility, and perception in driving. Perception **9**, 183–216 (1980)
9. S.I. Gerathewohl, Conspicuity of steady and flashing light signals: Variation of contrast. J. Opt. Soc. Am. **43**, 567–571 (1953)
10. J.J. Vos, A. Van Meeteren, Visual processes involved in seeing flashes, in *International Symposium of Imperial College of London: The Perception and Application of Flashing Lights*, (Adam Hilger Ltd., London, 1971), pp. 3–16
11. M.J. Allen, J. Strickland, A.J. Adams, Visibility of red, green, amber and white signal lights in a highway scene. Am. J. Optom. Arch. Am. Acad. Optom. **44**, 105–109 (1967)
12. M.J. Allen, *Vision and Highway Safety* (Chihon Book Co., Philadelphia, 1970), pp. 125–199
13. W.H. Lamm, Vehicle warning systems. Emergency **15**, 32–35 (1983)
14. L.R. Schwartz, Pre-hospital care, in *Principles and Practice of Emergency Medicine*, ed. by G. R. Schwartz et al., 2nd edn., (WB Saunders Co., Philadelphia, 1986), pp. 593–610
15. D.A. De Lorenzo, M.A. Eilers, Lights and siren: A review of emergency vehicle warning systems. Ann. Emerg. Med. **20**(12), 1331–1335 (1991)
16. R.D. Patterson, Auditory warning sounds in the work environment. Philos. Trans. R. Soc. Lond. **327**, 485–492 (1990)
17. S.C. Skeiber, R.L. Mason, R.C. Potter, Effectiveness of audible warning devices on emergency vehicles. Sound Vibrat. **12**, 14–17 (1978)., 20–22
18. S. Fidell, K.S. Pearsons, R. Bennett, Prediction of aural detectability of noise signals. Hum. Factors **16**, 373–383 (1974)
19. S.C. Skeiber, R.L. Mason, R.C. Potter, Effectiveness of Audible Devices on Emergency Vehicles (US Department of Transportation, National Highway Traffic Safety Administration, Washington, DC, 1977), Publication no. DOT-TSC-OST-77-3g
20. E.B. Hamida, H. Noura, W. Znaidi, Security of cooperative intelligent transport systems: Standards, threats analysis and cryptographic countermeasures. Electronics **4**, 380–423 (2015)
21. R. Barskar, M. Chawla, Vehicular ad hoc networks and its applications in diversified fields. Int. J. Comput. Appl. **123**(10), 7–11 (2015)
22. H.C.A. Van Tilborg, *Encyclopedia of Cryptography and Security* (Springer, New York, 2011). ISBN: 978-1-4419-5905-8
23. J. Ďurech, Bezpečnostné riešenia VANET siete pre riadenie inteligentných dopravných systémov, PhD thesis, University of Žilina, 2016
24. M. Raya, J.-P. Hubaux, Securing vehicular ad hoc networks. J. Comput. Security **15**, 39–68 (2007)
25. H.T. Cheng, H. Shan, W. Zhuang, Infotainment and Road Safety Service Support in Vehicular Networking: From a Communication Perspective, in *Mechanical Systems and Signal Processing*, vol. 25, (Elsevier, New York, 2011), pp. 2020–2038
26. S. Oh, M. Gruteser, D. Pompili, Coordination-free safety messages dissemination protocol for vehicular network. IEEE Trans. Veh. Technol. (99) (2012). https://doi.org/10.1109/TVT.2012.2197871
27. IEEE: 802.11p-2010—IEEE standard for information technology—Local and metropolitan area networks—Specific requirements—Part 11: Wireless LAN Medium Access Control (MAC) and Physical Layer (PHY) specifications amendment 6: Wireless access in vehicular environments, https://standards.ieee.org/standard/802_11p-2010.html. Last accessed 17 Apr 2019
28. 3GPP TS 36.300: Evolved universal terrestrial radio access (E-UTRA) and evolved universal terrestrial radio access network (E-UTRAN); Overall description; Stage 2

29. R. Molina-Masegosa, J. Gozalvez, M. Sepulcre, Configuration of the C-V2X mode 4 sidelink PC5 interface for vehicular communication, in *2018 14th International conference on mobile ad-hoc and sensor networks (MSN)*, Shenyang, China (2018), pp. 43–48
30. A. Morgado, K. Huq, S. Mumtaz, J. Rodriguez, A survey of 5G technologies: Regulatory, standardization and industrial perspectives. Digital Commun. Networks **4**(2), 87–97 (2018). ISSN: 2352-8648
31. D. Eckhoff, N. Sofra, R. German, A performance study of cooperative awareness in ETSI ITS G5 and IEEE wave, in *Proceedings of the 10th Annual Conference on Wireless On-Demand Network Systems and Services*, (IEEE, Piscataway, NJ, 2013), pp. 196–200
32. D. Krajzewicz, J. Erdmann, M. Behrisch, L. Bieker, Recent development and applications of SUMO—Simulation of Urban MObility. Int. J. Adv. Syst. Measur. **5**(3&4), 128–138 (2012)
33. OpenStreetMap, https://www.openstreetmap.org/. Last accessed 17 Apr 2019
34. Územný generel dopravy Mesta Žilina, https://www.enviroportal.sk/sk/eia/detail/uzemny-generel-dopravy-mesta-zilina. Last accessed 17 Apr 2019
35. OMNeT++ Discrete Event Simulator, https://omnetpp.org/. Last accessed 17 Apr 2019
36. European Commission. COM(2016) 766, A European strategy on Cooperative Intelligent Transport Systems, a milestone towards cooperative, connected and automated mobility, 30 Oct 2016, https://ec.europa.eu/energy/sites/ener/files/documents/1_en_act_part1_v5.pdf. Last accessed 17 Apr 2019
37. A. Buchenscheit, F. Schaub, F. Kargl, M. Weber, A VANET-based emergency vehicle warning system, in *2009 IEEE Vehicular Networking Conference (VNC)* (2009), pp. 1–8. https://doi.org/10.1109/VNC.2009.5416384
38. T. Petrov, V. Ďurček, M. Dado, K.E. Ambrosch, Study of performance of the vehicular ad hoc networks in dense network scenarios, in *Intelligent Transport Systems—From Research and Development to the Market Uptake*, ed. by T. Kováčiková, Ľ. Buzna, G. Pourhashem, G. Lugano, Y. Cornet, N. Lugano, (Springer International Publishing, Cham, 2018), pp. 85–89
39. T. Petrov, Simulation of routing protocols for emergency vehicle warning application in VANETs, in *22nd International Student Conference on Electrical Engineering POSTER*, (CVUT, Prague, 2018)
40. T. Petrov, M. Dado, K.E. Ambrosch, Computer modelling of cooperative intelligent transportation systems. Proc. Eng. **192**, 683–688 (2017). https://doi.org/10.1016/j.proeng.2017.06.118

The Specifics of the Retail Network and Consumer Shopping Behaviour in Selected Regional Towns of West Slovakia

Miroslava Trembošová, Alena Dubcová, Ľudmila Nagyová, and Dagmar Cagáňová

1 Introduction

The gradual adjustment of the economic policy of post-socialist economies towards the market has resulted in urban spatial structures in decentralization of retail functions, the modernization of traditional shopping malls, new synergies, new patterns of commercial supply, but also in the processes of filling "white spaces" with the youngest forms of stores—shopping centres. Dynamic restructuring of stores, construction of retail operations of large-format type creates the traction effect of other services relocation within city departments. Such a development can be an opportunity for urban zones in the form of their balanced development, as opposed to a possible dominant position of only one function/interest [1]. There is a multifaceted conflict of interest in these processes. Gaining retrospective control of evolving market processes is a common public sector challenge in spatial retail restructuring [1]. Rapidly changing technologies, accelerated cycles in the economy, precarious employment, a fragile level of the population purchasing power, etc., affect the nature of retail sales processes at various spatial and management levels. The owners and operators of shops are under constant pressure to attract customers,

M. Trembošová · A. Dubcová
Constantine the Philosopher University in Nitra, Nitra, Slovakia
e-mail: mtrembosova@ukf.sk; adubcova@ukf.sk; aldubcova@gmail.com

Ľ. Nagyová
Slovak University Agriculture in Nitra, Nitra, Slovakia
e-mail: ludmila.nagyova@uniag.sk

D. Cagáňová (✉)
Faculty of Materials Science and Technology in Trnava, Slovak University of Technology in Bratislava, Trnava, Slovakia
e-mail: dagmar.caganova@stuba.sk

D. Cagáňová et al. (eds.), *Advances in Industrial Internet of Things, Engineering and Management*, EAI/Springer Innovations in Communication and Computing,
https://doi.org/10.1007/978-3-030-69705-1_3

especially in anchored formats. Attractive advertising, quality of services, atmosphere of stores, environmental side of sales, etc. are steps that take into account the changing needs of customers.

Likewise, a "dark shop" that is not spatially anchored, with cloud attribute, needs to understand the changing demands of consumers. According to statistics, the online store provides a tenth of retail sales. It is evident that shopping is moving to online space faster than previously thought [2]. Various new forms are emerging, e.g. penetration of sales with an online store in the form of a 3D offer, beacon, click & collect, wearbless, Pygmalios Analytics software [3].

The prosperity of the online store also offers new applicable possibilities in stores. It is the identification of consumer habits and their implementation into the life of stores that has changed the retail network from static to dynamic, which is catching up with the implementation of various types of information technology.

The scientific monograph chapter presents the state and development of the spatial and functional configuration of three regional cities—Nitra, Trnava and Žilina retail network. At the same time, a view of visitors shopping behaviour is created according to the respondents in the most attractive business centres of these cities and the testing the respondents' demographic indicators influence on the types of their shopping behaviour is created.

The surveyed cities belong to the larger cities of Slovakia (60,000–80,000 inhabitants). They are located in the production region of western Slovakia. Their most important common feature is the location of the foreign corporations automotive industry focused on small cars the production and in the case of Jaguar Landrover luxury B-type vehicles. In all three cities, car manufacturers were built "on a green field" in peripheral urban areas (Trnava—PSA Peugot Citroën with production since 2006 in district-Trnava-East, 4500 employees, Nitra—Jaguar Land Rover Slovakia with production since 2018 in district Drážovce, 2000 employees, and in the village of Teplička nad Váhom adjacent to the city of Žilina—Kia Motors Slovakia with production since 2007, 3800 employees). All three cities as important centres of the Slovak universities are characterized by a high proportion of the students who significantly influence the structure of the retail network.

2 Theoretical Background

Business is an essential part of everyone existence. The retail has become not only a place to sell goods and services, but also a place where is spent free time [4]. In headquarters, retail stores are the most variable elements, they are lively, dynamic and spectral variable components, which change the spatial-functional structures of cities the fastest and most significantly [5]. Business activities, also known as market functions, have attracted the individuals to urban settlements for centuries. The last link in the retail channel is the store. The success of business in this sector is influenced by factors, the most important of which is the location and sales area [6]. Other authors, such as [7] point out the growing importance of other factors such as equipment and atmosphere of stores and offered assortment. In addition to a large number of such factors, non-commercial factors are also getting to the forefront, and the non-commercial use of leisure outlets affects the viability of retail network locations [8].

Despite the growing interest in the world to study the internal relationships between spatial models of retail localization in the urban environment and shopping behaviour [9], works with a complex solution of a comprehensive retail network consisting of large-scale and small-scale stores at the same time appear only sporadically.

In scientific geographical works on retail, the location factor of a place is emphasized. Salvaneschi [10], Jones and Simmons [11], Miller and O'Kelly [12] point out that not all new retail projects are captured. For this reason, especially in full cities with three or more large centres, the famous real estate saying, locality will apply. That is why the attention of many experts is directed to the study of the location and attractiveness of shops in relation to accessibility, whether on foot (especially in centralized parts of the city), by public transport or by car. Following the arrival of large multinational retail corporations in Slovakia, the investors and the public sector did not pay adequate attention to the aspect of ensuring sufficient buyers for newly established business operations such as shopping centres. On the one hand, the investors relied on the economic principle of "the hand of the market will shake it" and on the other hand on sufficient experience in building modern "large centres" in countries with developed market systems with different degrees of economic development, cultures, size, etc. Only the rapid depopulation of lively urban centres in the last decade of the twentieth century, the extinction of small shops, increasing security risks for visitors to multinational chains, the emergence of food deserts, etc., forced increased attention from experts in urbanism, geography, architecture and regional development.

Retail has been the key measure of urban vitality in many studies [13]. The concept of the "living centre" is a central theme of Hillier's theory of centrality, in which the pedestrian movement models using retail and other services in the city centre have a prominent position. Church and Murray [14] provide an overview of localization overlapping models and concepts not only for the retail sector [14]. At the same time, they point out the absence of a solution to the problem with the location of several devices, where the sale can also be cannibalized by another sale. According to [15], the development of retail in the city centre, in its suburbs or outside the city should be considered when assessing intra-city schemes. In this context, it is possible to use the classification of shopping centres according to their location [16], which distinguishes: (1) edge-of-centre; (2) out-of-centre; and (3) out-of-town and used by [17] on the example of Slovakia capital—Bratislava.

In recent decades, in many foreign cities but also in Slovakia, there has been a decline in central business districts and an increase in retail sales in peripheral areas, which has caused a fundamental shift in the place and way of shopping [18]. As a result of the large centre's construction and hypermarkets, there is a decrease in vacant space and there is a thickening of space in urban areas [19]. Nagy also drew attention to the need to revitalize traditional retail spaces. More and more experts are pointing to the need to revitalize city centres with retail services. Porta et al. [20] pointed out how little we know about the processes by which centres are created and maintained. Historically, it is clear that centres are not only growing and shrinking, but also shifting and diversifying.

Cities are trying to attract both consumers and investors to local retail centres, which often leads to competition. Therefore, retail development has become an arena of competition between regional cities and retail entrepreneurs. Cities and municipalities sometimes facilitate the planning and development of spaces with their specific use, which can lead to their fragmentation. The concept of a fragmented urban area is in fact borrowed from ecology, but it is not the only ecological concept adopted in the study of retail urban environments [21].

In the V4 countries, spatial and functional changes in retail were more dynamic than in the "traditional Western European" countries and entered the arena, especially of urban systems. As noted by [22], the transition of post-socialist countries to market economies has proved to be a challenging experience due to their absence and difficulty in building the appropriate political, legal, economic and trade structures needed for a free market economy.

In the Czech Republic [23] have identified two stages in the development of retail, the so-called "Atomization and internationalization of retail associated with its concentration, which has manifested itself in the Czech Republic since 1995." In Slovakia [24] based on 20 years of research (1992–2012) on the example of the regional city of Nitra confirmed three ongoing stages: (1) atomization, (2) consolidation, (3) concentration and initial manifestations of the saturation stage (in 2012). At the same time, they qualitatively name a new stage of demassification. The authors in Polish [25] as well as Hungarian [26] literature also dealt with the individual phases of retail development. On the example of St. Petersburg, Aksenov [27] and Aksenov et al. [28] have evaluated three factors: (1) stratification of shopping types, (2) development of certain retail facilities and (3) macro trends in the spatial organization of the urban economy tertiary sector and based on them presented six stages of retail development: (1) 1989–1996 early transformation; (2) 1996–1998 the first stage of administrative proceedings, (3) 1998–2001 second stage of stores and markets, (4) 2001–2002 the post-transformation large-format redistributions, (5) 2002–2007 the phase of large-format stores dominance, (6) 2007–2016 post-transformation stage development of neighbourhood stores, small store formats, development of new non-traditional locations and alternative food sales networks.

The development of retail in the European countries is significantly influenced by regulation aimed at strengthening and helping small local independent retailers. Restriction measures in the retail sector began in OECD countries in response to globalization and the "cannibalistic impact" of shopping centres on local stores [29–31]. Even France, the birthplace of the world's first hypermarket (Carrefour, 1963), entered into the restrictions as early as the early 1970s [32]. In the UK, the environmental restrictions on retail regulation began to be applied on a larger scale in the last decade of the twentieth century. They have analysed the effects of the English restrictions for the period 1997 to 2003 [33]. They have examined the effects of helping smaller retailers in city centres and restricting the entry of large chains [34]. Schivardi and Viviano [34] point out that strong regulation in retail also creates a negative impact on unemployment, as the retail sector employs about 10% of the economically active population in all market economies. The development of retail and the consideration of the regulatory measures in the countries adoption in Central and Eastern Europe CEE in various forms, proposals and structures are dealt with by several authors, e.g. [15, 35–39]. In addition, there is still a debate about the extent to which the regulation affects the social, economic and environmental habits of consumers [40–45].

Retailers are constantly facing a dynamic and competitive retail environment. With increasing globalization, market saturation and increased competitiveness through mergers and acquisitions, they are looking for competitive advantages, e.g. better customer relationship management through the creation of customer databases. Shopping is one of the basic recurring human-space-time activities [46]. Due to technological progress, the time fund of the population is changing and new shopping opportunities are changing. At the same time, purchasing time is shortening [47]. According to their research, this phenomenon is related to the work of women away from home, the shortening of cooking time, but also the subsequent increased demand for processed food, and the fast way of shopping. Shopping also reflects the changed technological scene and radically changed consumer behaviour, which at the same time forces retailers to rethink their business models, retail formats and retail offerings [48]. According to [49] retailers can effectively predict consumer behaviour, design attractive offers better targeted at their customers and develop new tools to attract customers with specific targeted consumer data /megadata/.

These large data can initiate beneficial cyclical processes of customer consumption and engagement, which in turn lead to increased profitability [50]. Trembošová and Tremboš [51] write about targeted information and an offer directly tailored for customers in the stage of retail demasification in Slovakia, as well as other foreign authors, e.g. [49]. A large group of authors deals with the consumer behaviour classifications. The customer typologies can be divided according to demographic, psychological, economic aspects (gender, mode of transport, frequency of shopping, types of goods purchased, the location where shopping takes place, shopping behaviour, context, etc.). E.g. typology [52] distinguishes four segments of visitors to (Turkish) shopping centres—serious, recreational, pragmatic and enthusiastic consumers. Alavi et al. [53] present a classification of four styles according to consumer decision-making (Malaysian customers)—hedonistic, fashionable, useful and impulsive.

Consumer behaviour strategies are influenced by the economic or political environment. E.g. economic crisis, already at the beginning of its operation fundamentally changes the income of the population, which consequently manifests in reduced purchasing power of goods in retail. These aspects were studied, e.g. [54–59] and others. Their conclusions showed that during this period, consumers moved significantly to discounted goods, but also increased the use of the Internet as well as preparation for shopping. The growing sales of discounted retail companies confirmed that the economic crisis has significantly conditioned changes in consumer behaviour in retail. The same conclusions were reached [24], whose results of questionnaire surveys from the environment of Nitra city (2008 and 2012) confirmed the shift from pre-crisis modern behaviour (respondents influenced by shopping, demanding and mobile), to the traditional crisis in 2012 (when they were more inclined to a frugal, loyal and cautious subtype).

3 Research, Methodology, Data, and Research Area

The input primary data for the needs of a spatial change and functional differentiation of the retail network detailed analysis in three regional western Slovakia cities, Trnava, Nitra and Žilina, as well as typologies of shopping behaviour were based on the field research and passportization of their retail network. The collection of data used in this study was carried out in the cities—Nitra (1992, 2002, 2005, 2008, 2010, 2015, 2017, 2018, 2019), Trnava (2002, 2015) and Žilina (2015) on the district level. The most representative results were obtained for the city of Nitra due to the longest ongoing research (1992–2019). The mapping of the retail network included small-area as well as large-area stores with different assortment, offers as well as different nature of goods (new, old, used goods). The location, the size of the sales area as well as the assortment structure of individual stores were monitored. The size and level of retail equipment were determined on the basis of a comparison of the area parameter (sales area per 1000 inhabitants—PAFS) with the European standard 1000 m^2/1000 inhabitants. MO equipment levels (local, urban, regional and superregional) were determined. The development of the retail network was monitored in individual cities. Based on the spatio-temporal typology, four stages of development were identified—the stage of atomization, consolidation, concentration and saturation with emphasis on the model area of the city of Nitra. Due to the spatio-temporal changes that occurred in this sector of the economy after the change of political and social conditions, the transformation from a monocentric to a polycentric model of spatial organization of retail trade was verified.

The attractiveness of shopping centres and the image of consumer behaviour of their visitors was created on the basis of soft data obtained in a behavioural survey in the regional cities of Nitra, Trnava and Žilina. The survey was conducted by random selection of respondents from places outside shopping centres in 2015. The basic tool of the survey was a questionnaire, from which the characteristics of respondents, perception of location of business entities and attractiveness of the shopping centre were selected for the given contribution. In attractive resp. in popular shopping centres, the shopping behaviour of respondents—customers—was studied. The research included three shopping centres in each city.

The research component consists of 1246 respondents, of which in Nitra (570), Trnava (347) and Žilina (329). According to the respondents, the most attractive shopping centres of these regional cities are also the subject of the research interest of this paper. In Nitra they include OC Galéria (37% of the research component of the city's respondents), OC Mlyny (31%) and OC Centro (32%), in Trnava—OC City Aréna (56%), OC Galéria (23%) and ZOC Max (25%) and in Žilina—OC Aupark (58% of visitors) and OC Mirage (33%) and ZOC Max (9%).

Selected shopping centres are the subject of a study of respondents' shopping behaviour. In addition to individual scientists, scientific teams, the issue of purchasing behaviour is dealt with by various specialized agencies, focused surveys in various areas of economic life, including retail. From them, we selected the Shopper Typology Media Behaviour methodology of Incoma Research and Gfk Praha from 2003 [59]. The methodology is focused on the types of shopping behaviour and their characteristics, which are based on the forms, motivation and method of shopping. It consists of two basic types and several subtypes of shopping orientation: (1) traditional (careful or conservative, frugal, loyal and unpretentious), (2) modern (influential, demanding and mobile or pragmatist). The issue of shopping behaviour is very complex and therefore we allowed respondents in the survey, according to their own discretion and the presented characteristics, to be included in the individual subtypes of consumers.

Subsequently, we tested whether shopping behaviour is depended on demographic characteristics (gender, age groups, education, economic activity, household size, and average household income). If there was an addiction, we monitored its intensity. We performed the testing using the test of square contingency of qualitative features, where we tested χ^2 at the significance level p with a value of 0.05. If the value of the calculated χ^2 is lower than the critical value, the two characters are independent, the dependence does not exist. If the value of the calculated χ^2 is lower than the critical value, the two characters are independent, the dependence does not exist. Otherwise, there is a dependency. The intensity of the relationships between the characteristics of the respondents and the types of shopping behaviour was determined on the basis of the coefficient of consistency C and verified using the Cramer's test V. The testing was processed in the SPSS program.

Pearson's contingency coefficient expresses the degree of dependence between two qualitative features in a contingency table and is determined by the relation:

$$C = \sqrt{\frac{X^2}{X^2 + n}}$$

where χ^2 is the Pearson test statistic (square contingency test) Enter the equation.

$$n = \sum_{i=1}^{r} \sum_{j=1}^{s} n_{ij}$$

The more C acquires values approaching 1, the higher the dependence of both qualitative features. Conversely, C values approaching 0 indicate very low to no dependence. The Cramer V coefficient represents the most appropriate degree of association between two qualitatively variable ones and is a modified version of the correlation coefficient. It is used for tables of variables larger than 2 × 2, which is true in our case. Cramer's V coefficient is expressed:

$$V = \sqrt{\frac{X^2}{n \cdot \min(m-1, k-1)}}$$

where χ^2 is Pearson's test statistic, n is the number of variables in the table, m is the number of columns and k is the number of rows in the table. We calculate Pearson's test statistic χ^2 based on the relation:

$$\chi^2 = \sum_{p=1}^{q} \frac{\left(f_{ep} - f_{op}\right)^2}{f_{op}}$$

where f are the frequencies of the variables arranged in q classes, f_{ep} are the empirical frequencies and f_{op} the expected frequencies.

4 Survey Results

According to [60], one of the visible different characteristics of Central European post-socialist cities from Western European capitalist cities is the population density gradient. As they write, in contrast to the gradually decreasing density from the centre to the periphery in market-oriented cities, post-socialist cities usually have "camelbacks" due to the "cavity" of the transition belt and the "top" of the outer zone of cities dominated by large housing estates. This situation is also significant for the researched regional cities. Settlement forms in the city of Nitra are represented by the housing estate/town district Klokočina (22,000 inhabitants), in the town Trnava by the settlement Družba (approx. 14,000 inhabitants) in the town district Trnava East and in the town Žilina by the settlement/town district Vlčince (approx. 18,000). This picture of the settlement structure of the city is the result of the construction of prefabricated housing estates in all three cities during the socialist period.

Despite the fact that Trnava, Nitra and Žilina have the same position in the settlement structure of Slovakia, their development after the transformation of the economy from a centrally managed to a market one at the end of the 1990s was different. Outwardly manifested, e.g. in the number of small and large stores, a change in the organizational structure of retail facilities, especially in the size of the sales area, in various types of shopping centres, assortment structure, etc. In terms of the number of stores between cities, Žilina has a significant position, surpassing Trnava and Nitra in numbers. The situation is different in the sales area, where Nitra is primary and Trnava has the smallest sales area (Table 1). The larger sales area

Table 1 Basic parameters of the retail network of regional cities Nitra, Trnava a Žilina in 2015

City	A	B	C	D	E	F	G	H
Nitra	80,161	871	160,919	184.7	2007	92.6	19	4
Trnava	64,439	769	120,559	157	1870	83.8	25	5
Žilina	83,591	952	120,623	126.7	1443	88	16	4

Note: A—number of inhabitants, B—number of stores, C—sales area, D—average size of store, E—PAFS the ratio of the population (in thousands) to the admissible floor space, F—the ratio of the population (individual) and number of stores, G—large-scale stores, H—shopping centres
Source: Authors' survey

Table 2 An overview of shopping centres in the city of Nitra, Trnava and Žilina in 2015

City	A	B	C	D	E	F
Nitra	Centro	Chrenová	2006	90	23,982	764
	Max	Chrenová	2006	75	16,487	450
	Galéria	Staré Mesto	2008	43	27,440	686
	Mlyny	Staré Mesto	2009	139	32,500	1100
	Nitra town		–	*347*	*100,409*	*3000*
Trnava	Max	Trnava-Stred	2004	37	10,000	126
	Galéria	Trnava Sever	2006	43	15,072	845
	Arkadia	Trnava-Sever	2006	28	8000	
	Mall Trnava	Trnava-Juh	2008	30	30,000	1285
	City Aréna	Trnava-Stred	2015	68	20,000	950
	Trnava town		–	*206*	*83,072*	*3206*
Žilina	Atrium Dubeň	Vlčince	2001	55	22,600	1080
	Max	Solinky	2009	43	25,431	490
	Aupark	Staré Mesto	2010	86	24,000	850
	Mirage	Staré Mesto	2010	50	25,671	388
	Žilina town		–	*234*	*97,702*	*2808*

Note: A—name shopping centre, B—city boroughs name, C—opening year, D—number of stores, E—gross leasable area in m^2, F—Number of parking lots
Sources: [61, 62] and authors' survey

in Nitra is a reflection of the larger number of large stores in the suburbs and the reduction of the commercial zone in the central part of the city, which was caused by the construction of OC Mlyny in the centre, where the retail stores from the centre were concentrated.

In the period from 2001 to 2015, up to 13 shopping centres were built in the surveyed cities. They started operating in the city of Nitra in 2006 and in 2015 there were four shopping centres (NC), which had a total gross leasable area (HPP) of 100,409 m^2 (Table 2). According to the location, three of them NC Centro, Max and Galéria belong to the "out-of-centre" type according to the classification [16], NC Mlyny to the "edge-of-centre" type. According to the stratification [23], these shopping centres can be included in the "second" (NC Centro and Galéria) and "third" (NC Max and Mlyny) generation of shopping centres. In Trnava, five shopping centres with a total gross leasable area (HPP) of 83,072 m^2 are available for the needs of the population (Table 2). Of these, the OC City Arena (opened in 2015) is unique in Central Europe due to its connection with the football stadium. It belongs to the "edge-of-centre" type and is a "third generation" shopping centre. The Max, Galéria and Arkadia centres belong to the "out-of-centre" type and, depending on the spatial character, it is a location of mixed type 1 and 2, which is characterized by spatial proximity to the customers' home and to competing companies. The Mall Trnava is from the group of "second-generation" edge-of-town shopping centres. In Žilina, four shopping centres with a total gross leasable area (HPP) of 82,700 m^2 (Table 2) have been in operation since 2001, of which Aupark and Mirage are "edge-of-centre" and two—Atrium Dubeň and Max type " out-of-centre ". All belong to "second generation" shopping malls.

4.1 Development Stages of the Retail Network After 1989

In the transformation period, the original state monopoly was replaced by a free-market mechanism. This first period was marked as a stage by foreign authors [23, 63, 64], but also by domestic ones [62, 65] designated as the atomization stage and the second as concentration stage. In none of the transforming economies of Central Europe [23, 64, 66, 67] has been confirmed consolidation stage only in the Slovak Republic. It is the result of the specific development of the Slovak economy, which after a period of weak growth experienced a significant boom in 2004–2008, which was reflected in the growth of household incomes, their purchasing power and the subsequent increased interest of foreign owners of large retail chains in the Slovak market.

The individual stages of the development of the retail network were studied to varying depths (Table 3) in post-socialist cities. In the environment of the Czech Republic, Poland and Hungary, the stage of concentration began relatively early, in 1996, and in Slovakia until 2004. This delayed process of concentration of the retail network is not over today, although we can observe signs of the next stage of development. Its development is a reflection of the development of the information society. The starting point may be the idea of the so-called fourth developmental wave of civilization [68], which is characterized by several features that can be embedded in the concept of demasification. It is characterized by lower or no sales area requirements [51].

The saturation stage was confirmed only in the city of Nitra. It occurs as standard at PAFS 2000 m^2/1000 inhabitants. It is characterized by some important features: (1) the growth of the sales area stops, there may be a decrease, (2) competition intensifies, (3) it is more difficult to acquire new tenants of commercial premises, (4) rents are declining, (5) part of the OC capacity is unused, (6) the emergence of new small-scale sales of regional products, (7) the emergence of alternative forms of sales, e.g. unpackaged shops, farmers' markets, box sales, and yard sales.

Table 3 Comparison of spatio-temporal phases of transformation and post-transformation processes in the spatial organization of retail of large cities in the V4 countries and Russia

Slovakia		Czechia, Poland, Hungary		Russia	
Period	Stage and characteristics of specific retail forms	Period	Stage and characteristics of specific retail forms	Period	Stage and characteristics of specific retail forms
1989–1996	*Atomization*	1989–1996	*Atomization*	1989–1996	*Early transformation*
	Disintegration, atomization		Disintegration—Atomization,		Mobile retail forms
	Retail network concentration		Retail network concentration		Kiosks
	Expansive growth of operations		Expansive growth of operations		Trade areas
	The importance of the city centre has declined				
1997–2003	*Consolidation*	1996–2015	*Concentration*	1996–1998	*First stage of the administrative reform*
	Minimal changes in indicators		The arrival of transnational chains		Transition from kiosks and trade areas to pavilions
	New stores located outside traditional retail zones		Construction of shopping centres		And open-air markets
	Low purchasing power of the population,		Increasing purchasing power		
	High unemployment rate		Relatively low unemployment rate	1998–2001	*Stage of pavilions and markets*
					Pavilions
					Open-air markets
					Stalls

(continued)

Table 3 (continued)

Slovakia		Czechia, Poland, Hungary		Russia	
Period	Stage and characteristics of specific retail forms	Period	Stage and characteristics of specific retail forms	Period	Stage and characteristics of specific retail forms
2004–2009	*Concentration*			2001–2002	*Stage of displacement of specific transformation forms*
	The arrival of transnational chains				New shopping malls
	Construction of shopping centres				Permanent markets
	The arrival of transnational chains				Kiosk chains
	Construction of shopping centres				
	Increasing purchasing power				
	Relatively low unemployment rate				
	Demasifikáca			2002–2007	*Stage of dominance of major permanent retail forms*
	The disintegration of the mass market into differentiated segments				Large and major retail forms start to absolutely dominate in turnover
	The emergence of new forms (catalogue sales, home teleshopping systems, e-commerce, direct mailing and other diverse channels and create a demasified market)				
					Temporary and illegal outlets (stalls, mobile)

2010–2014	*Saturation*			2007–2015	*Stage of neighbourhood stores*
	Post-crisis period				Neighbourhood stores
	occurs after exceeding the PAFS 2000 m^2/1000 inhabitants				Shop-in-shop
	Competition is intensifying				
	Disruption of operations in shopping centres				
	Rents are falling				
	Gradual cleaning of the retail scene				

Sources: Adjusted according to [23, 26, 27, 61, 62, 64]

4.2 *Spatial Configurations of the Retail Network of Selected Regional Cities According to PAFS*

One of the maim criteria for assessing the area's retail network equipment level is the sales area in relation to the number of inhabitants. Based on foreign experience, the optimized equipment of cities with more than 50,000 inhabitants is at the level of **1200–1400 m^2/1000 inhabitants** area parameter. In Central Europe, the value of 1000 m^2/1000 inhabitants has been used for the PAFS average for a long time [24, 27]. In Table 4 are shown the average values of PAFS of the examined regional cities (Nitra, Trnava and Žilina), which we compared with selected regional cities in the Czech Republic.

Based on the data on sales areas and the application of the calculation of the area parameter at the level of urban districts, a four-level hierarchy of urban districts was created, based on the PAFS average for EU cities and thus the value is 1000 m^2/1000 inhabitants. This value represents the lower limit of the city level. Other levels are (1) local: up to 250 m^2/1000 inhabitants, (2) urban: 251–1500 m^2/1000 inhabitants, (3) regional: 1501–4000 m^2/1000 inhabitants and supra-regional hierarchical level: 4001 and more m^2/1000 inhabitants. The mentioned hierarchization was used for the area parameter of the MO network of the city of Nitra [5, 62], Trnava [61] and Žilina [63].

The arrival of multinational chains, which brought large-scale store formats with an area of over 2500 m^2, had a dominant influence on the configuration and spatial-functional organization of the retail network of Slovak cities. Their tumultuous arrival in the city of Nitra has been evident since 2004 (Table 5). Almost all foreign chains in the city except Tesco—a department store (closed at the time and in restructuring since 2017) were built on a green field.

During the economic crisis and in the post-crisis period of years (in the years 2009–2015), the sharp construction of large-area stores slowed down in the city, and even the average area parameter of Nitra decreased. Nevertheless, the indicator remained high (2007 m^2/1000 inhabitants in 2015). During this period, several smaller stores in individual city districts ceased to exist. The construction of the

Table 4 The ratio of the population (in thousands) to the admissible floor space retail equipment in selected regional cities of the Slovak Republic and the Czech Republic

City	PAFS the ratio of the population (in thousands) to the admissible floor space					
Year	1992	1998	2000	2002	2010	2015
Nitra	565	794	780	824	2223	2007
Trnava	x	x	x	693	x	1870
Žilina	x	x	x	x	x	1443
Olomouc	x	846	x	x	x	x
Brno	x	x	1100	1780	x	2300
Zlín	x	x	960	x	x	x

Source: [62]

Table 5 Expansion of large-scale foreign retail chains in the city of Nitra in the years 2000–2019

Chain	**200...**										**201...**									
	0	**1**	**2**	**3**	**4**	**5**	**6**	**7**	**8**	**9**	**0**	**1**	**2**	**3**	**4**	**5**	**6**	**7**	**8**	**9**
Tesco	■	■	■	■	■	■	■	■	■	■	■	■	■	■	■	■	■	■	■	■
Baumax	■	■	■	■	■	■	■	■	■	■	■	■	■	■	■					
Metro	■	■	■	■	■	■	■	■	■	■	■	■	■	■	■	■	■	■	■	■
Billa		■	■	■	■	■	■	■	■	■	■	■	■	■	■	■	■	■	■	■
DM drogéria					■	■	■	■	■	■	■	■	■	■	■	■	■	■	■	■
Lidl					■	■	■	■	■	■	■	■	■	■	■	■	■	■	■	■
Kaufland						■	■	■	■	■	■	■	■	■	■	■	■	■	■	■
NAY Elektro						■	■	■	■	■	■	■	■	■	■	■	■	■	■	■
Hypernova							■	■	■	■	■	■	■	■	■	■	■			
Asko								■	■	■	■	■	■	■	■	■	■	■	■	■
Mountfield								■	■	■	■	■	■	■	■	■	■	■	■	■
Kik									■	■	■	■	■	■	■	■	■	■	■	■
JYSK									■	■	■	■	■	■	■	■	■	■	■	■
Deichmann									■	■	■	■	■	■	■	■	■	■	■	■
Takko									■	■	■	■	■	■	■	■	■	■	■	■
KIKA									■	■	■	■	■	■	■	■	■	■	■	■
MerkuryMark											■	■	■	■	■	■	■	■	■	■
Kärcher															■	■	■	■	■	■
Decodom															■	■	■	■	■	■
OBI																■	■	■	■	■
Möbelix																■	■	■	■	■
Total	3	4	4	4	6	8	9	11	16	16	17	17	17	17	19	20	20	19	19	19

Source: [62], internal sources of field research 2017, 2018, 2019

largest NC Mlyny (2009) in the city centre in the Old Town was reflected in the change of facilities from the regional to the superregional level. City districts with wholesale shopping centres Chrenová, Klokočina and Mlynarce, Horné Krškany acquired a regional level, although in 2015 a supraregional level was identified in them.

During the period 1992–2019, three chains left the Nitra retail scene—the English chain K Mart (1992), the German hobby market chain Baumax (2015) and the Czech chain of the AHOLD Hypernova group (2017). Based on the research of the sales network in the period 1992–2019, it is possible to say that the increase in sales area was dynamic, which was also reflected in the growth of the values of the city average. In 1992, three levels of retail network equipment were identified in the city of Nitra—local, urban and regional, by 2002 they were transformed into four (local, urban and regional and superregional), which were basically maintained until 2019, although with minor spatial changes (Fig. 1). This development of the area parameter at the level of city districts is characterized by **the change of the monocentric model** of the city of Nitra to a **polycentric model** in 2015 where

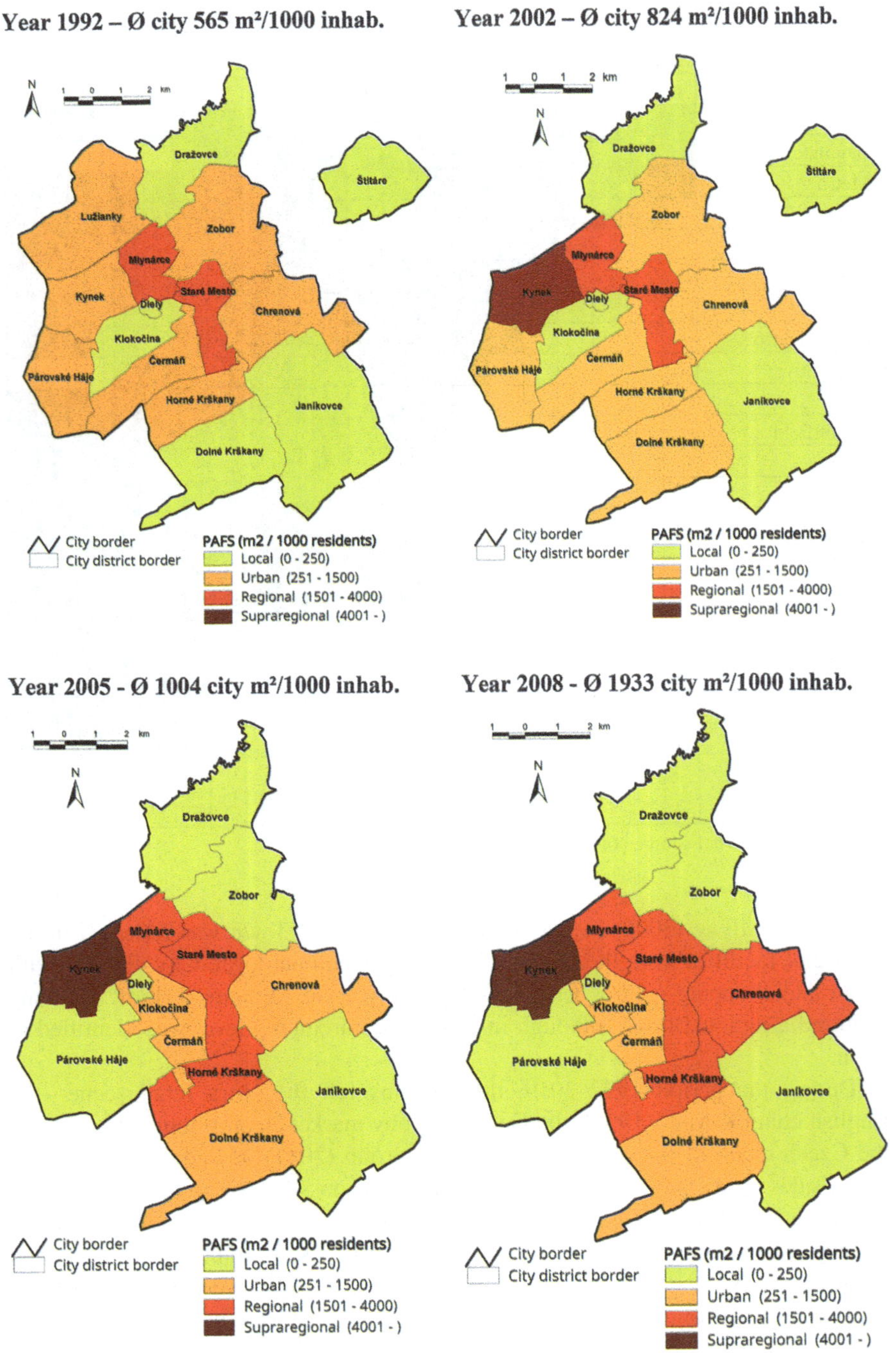

Fig. 1 PAFS (the population (in thousands) to the admissible floor space) according to the city districts of the city of Nitra in 1992, 2002, 2010, 2015, 2017 and 2019. Note: a part of the

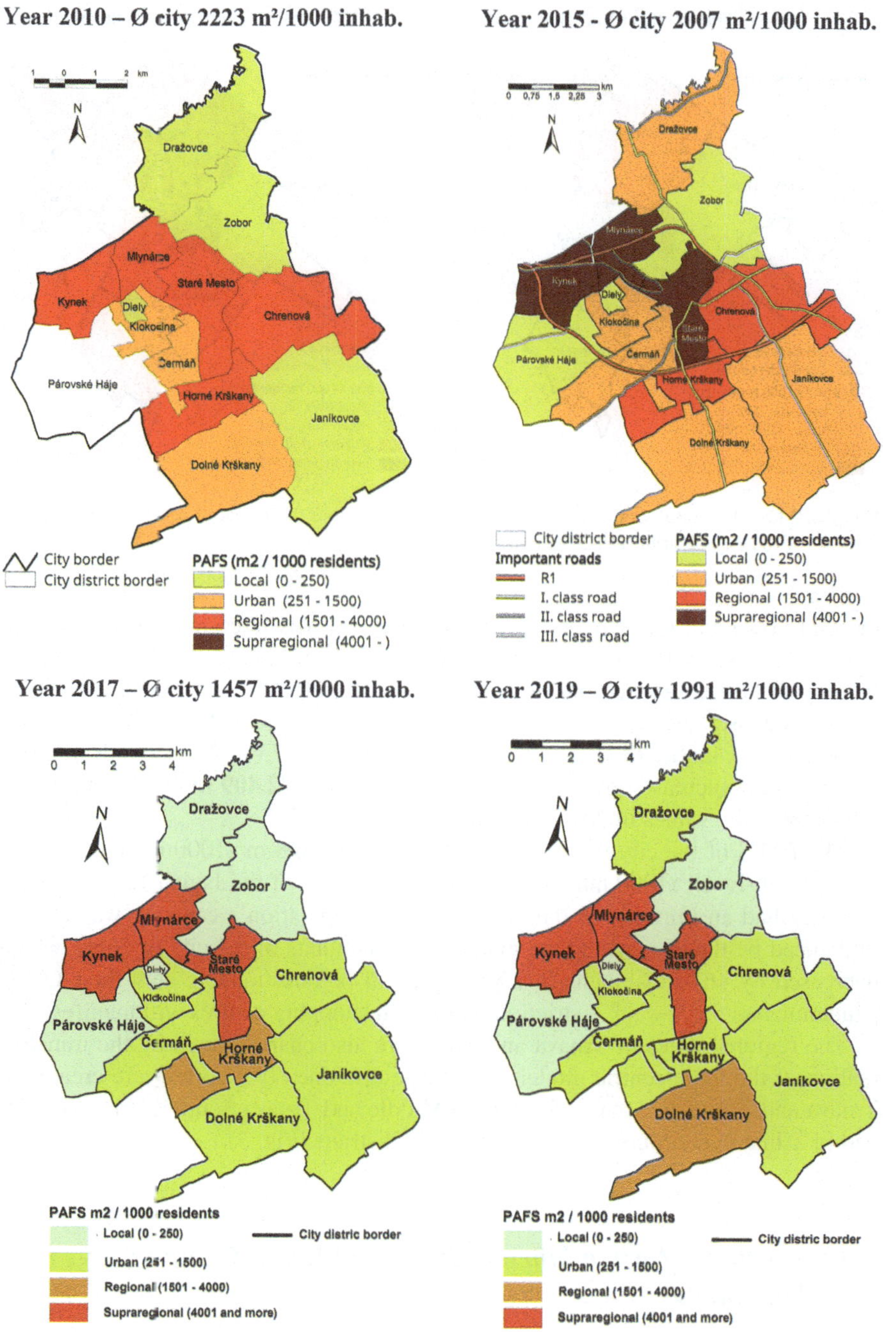

Fig. 1 (continued) town in 1992 is also the town part Lužianky, which was separated in 1993, Štitáre in 2002 and the adjustment of the borders between the town districts Parovské Háje and Čermáň in 2012th. (*Sources*: [62], the authors survey, 2015, 2017, 2019)

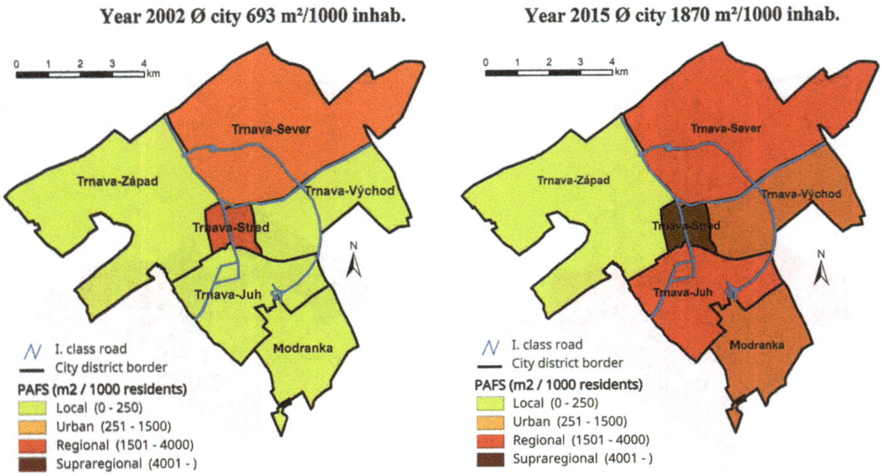

Fig. 2 PAFS (the population (in thousands) to the admissible floor space) according to the city districts of the city of Trnava in 2002 and in 2015. (*Source*: [61])

the Old Town and the two city districts Kynek and Mlynárce reach superregional importance. According to the latest data from field research (2017 and 2019) in the city of Nitra, after the purification stage (2010–2014), the concentration stage manifested itself again. The evidence is confirmed by data from 2017 to 2019. Although the number of retail entities decreased from 773 in 2017 to 748 in 2019, the sales area increased from 112,177 m^2 in 2017 to **152,409 m^2** and is gradually returning to the values in 2015, i.e. 160,919 m^2.

The PAFS of the city of Nitra increased from 1456.8 m^2/1000 inhabitants in 2 years to 1991.4 m^2/1000 inhabitants. The construction of the Jaguar Landrover car complex had an impact, but the increase in the population's capacity was greatly influenced by the decrease in the number of inhabitants of the city of Nitra, which decreased by 467 inhabitants in 2 years. It will be questionable how the COVID pandemic and the 9-week restriction on sales in shopping centres will manifest.

The regional cities of Trnava and Žilina are also characterized by the transformation of the monocentric model to a polycentric one. Two cores were created in Trnava and Žilina. In Trnava, the Trnava Middle and Trnava South districts (Fig. 2) and in Žilina in the Staré Mesto and Rosinky districts (Fig. 3).

4.3 Specifics of Attendance of Preferred Retail Centres: Popularity

Shopping is one of the basic activities without which we cannot imagine human life today. It becomes part of our lifestyle. Shopping from small isolated stores

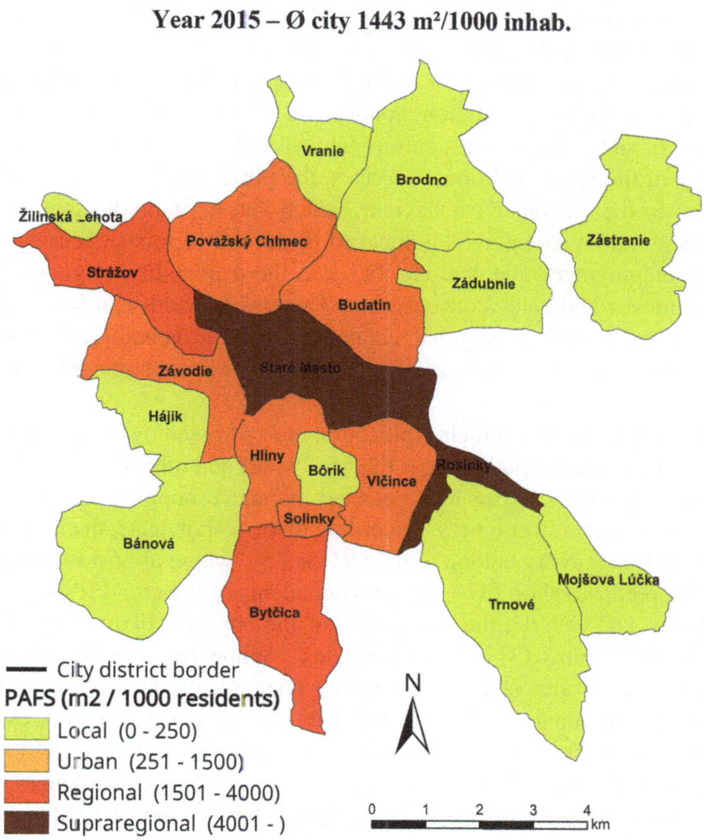

Fig. 3 PAFS (the population (in thousands) to the admissible floor space) according to the city districts of the city of Žilina in 2015. (*Source*: [69])

is now moved to large shopping centres, which provide a wide range of goods in branded stores but also have large-scale grocery stores. Shopping centres have an increasingly strong position in Slovak retail, they partichipate to a large extent in retail turnover and their number is constantly increasing [45].

In addition to shopping, customers also use other service services of shopping centres such as postal, catering, entertainment, sports, cultural, etc. Shopping centres are becoming more and more social places, meeting with friends, family, while using free-time they fulfil the function of "shopping centre socialites". They create space for maximizing the length of the customer's stay in the shopping centre and the associated increase in average expenses for the purchase of goods and services [45]. Through the offered attractions and services, they create a recreational potential for the development of urban tourism [70].

Consumer behaviour is according to [71] the object of a process study that covers a large number of causes of choice of purchase, use of products, by individuals or groups in order to meet their desires and needs. Križan et al. [71] confirmed that different individuals exhibit different shopping behaviours. A wide range of different subjective attributes enter into their shopping behaviour, such as customer mood, customs, traditions, religion, atmosphere of the environment, mix of shops, willingness of staff, image, sales culture, etc. as well as objectively, such as price, availability of the store, additional services, the possibility of parking, etc., which is reflected in the customer's behaviour, which can be unpredictable, sometimes illogical. The rate of shopping is influenced by several factors such as age, gender, education, employment, income, etc. but also the availability of goods, lifestyle, marketing moves and sales concepts, etc. Their wide width can be grouped into several basic areas such as personal, economic, social, psychological, and cultural. These factors make it possible to segment consumers while adapting their marketing mix.

Consumer behaviour and their typification are addressed by several foreign [72–78] but also by domestic authors [61, 62, 65, 79–82] and others.

The vast majority of buyers have selected (popular) shopping centres, which for various reasons prefer over others. According to the respondents, the most attractive shopping centres in Nitra belong to the OC Galéria, where almost two-fifths of the respondents prefer to shop (37% of the respondents of the city's research branch). In addition to OC Galéria, the inhabitants of the city also like to buy OC Mlyny (31%) and OC Centro (31% of respondents). While in Nitra the popularity of shopping centres is almost balanced among respondents, in Trnava the OC City Arena strongly dominates (56%). In OC City Aréna, twice as many respondents make their purchases as in other shopping centres—OC Galéria (23%) and ZOC Max (25%). In Žilina, the most visited shopping centres include Aupark (58% of visitors), which in popularity exceeds OC Mirage (33%) and ZOC Max (9%). The basic characteristics of the respondents according to individual department stores are documented in Fig. 4.

The attractiveness of shopping centres is increased by their location as well as convenient accessibility by means of transport. Nagyová [6] points out that the choice of location and the correct location of the shopping centre, its equipment as well as the range of goods offered are one of the most important decisions and the most important factor in the success of a retail entity to attract customers. According to the respondents (Table 6), the shopping centres (OC Aupark, OC Mirage, OC Mlyny, OC City Aréna) are suitably located as "edge-of-centres". It is their location in the central parts of the city or on their outskirts that increases their attractiveness. These shopping centres are located near strategic transport hubs or bus stops. In addition, they solve a significant problem of city centres—the lack of parking spaces with their large parking spaces. The location of the Galéria shopping centres in Trnava and Nitra as well as ZOC Max in Žilina, which have an "out-of-centre" location, are also perceived by the respondents as highly positive (Table 6). Surprisingly lower ratings for the location of shopping centres have shopping centres in Nitra (66% to 53%), which indicates that, convenient location, easy accessibility, plenty of parking spaces for customers over time, can be replaced by

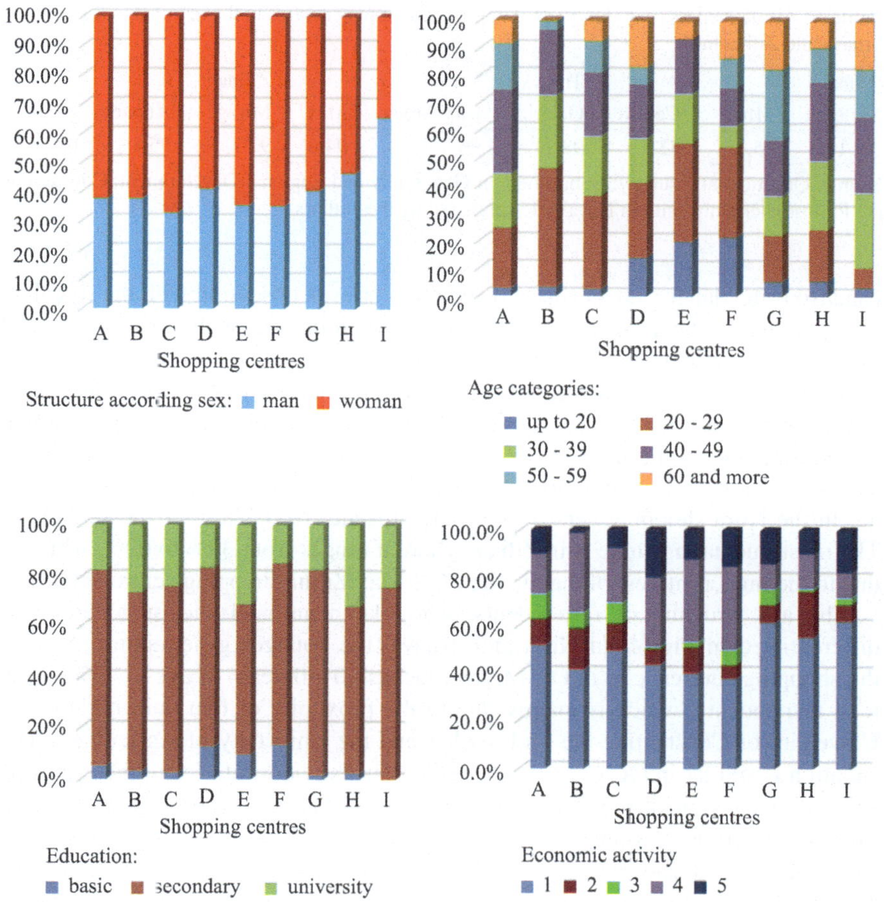

Fig. 4 Basic characteristics of respondents of department stores in the regional cities of Nitra, Trnava and Žilina. Note: A—OC Galéria Nitra, B—OC Mlyny Nitra, C—OC Centro Nitra, D—OC Galéria Trnava, E—OC City Aréna Trnava, F—OC ZOC Max Trnava, G—OC Mirage Žilina, H—OC Aupark Žilina, I—OC ZOC Max Žilina. Economic activity: 1—employee, 2—businessman, 3—unemployed, 4—student, 5—pensioner. (*Source*: Questionnaire survey: Department of Geography and Regional Development, Constantine the Philosopher University in Nitra 2014, 2015, prepared by the authors)

more attractive factors such as mix of branded stores, non-commercial services resp. use of free time, etc.

In the research component of attractive shopping centres in cities, women have a more prominent position than men. Women like to shop, they not only shop for themselves, but they also like to give gifts to their loved ones and friends. They often see in shopping fun and a great way to relax, spend free time, self-realization but also creativity. They mostly do not shop alone, but with girlfriends, colleagues or family. By shopping, they improve their image, their position in society. Shopping

Table 6 Attractiveness (popularity) of shopping in shopping centres in the regional cities of Nitra, Trnava and Žilina in %

Nitra			Trnava			Žilina		
Galéria	Mlyny	Centro	Galéria	City Aréna	Max	Mirage	Aupark	Max
57.1	65.7	53.3	98.7	96.2	94.6	92.7	95.8	82.8

Source: Questionnaire survey: Department of Geography and Regional Development, Constantine the Philosopher University in Nitra 2015, prepared by the authors

can also bring danger, it can slip into addiction or even obsession with the so-called "Shopoholic" to the desire to buy any things, even if the person does not really need them. This addiction can develop into oniomania.

When shopping, men are considered more rational, they buy mostly less, they buy what they absolutely need or according to their hobbies, e.g. food, electrical engineering, etc. In the study sample, there were 1.6 women per one shopping man. The ratio of respondents in individual shopping centres of cities is expressed in Fig. 4a. In the Nitra shopping centres, the range reaches 62–67%, in Trnava 59–65%. The most fluctuating share is in Žilina, where it ranges from 39% to 60%, which is due to the higher interest of men in the ZOC Max Žilina shopping centre (61%).

The age structure of respondents, who like mentioned shopping centres is differentiated in Fig. 4. In Nitra and Trnava, the younger generation dominates the shopping centres, e.g. in OC Mlyny reaches 20–30 years of age to 43% of the total respondents, which is mainly due to the proximity of two universities—the University of Constantine the Philosopher and the University of Agriculture. The situation is similar in OC Centro, where the university dormitories of these universities are located next to it. Their share of respondents reached one-third (33%). In Trnava, this group of young people has a third share in two shopping centres (OC City Aréna—32% and ZOC Max—32%). In Žilina, the OC Aupark (18%) and OC Mirage (16%) centres are particularly attractive for young respondents. Shopping centres provide young people not only with a mix of shops with branded goods, but especially with spending their free time, meeting with friends, doing sports or relax such as in OC City Arena, which is connected to the sports stadium. The attractiveness of these centres is mainly influenced by the proximity of universities and high schools.

Young people create generation Y (20–40 years) and Z (up to 20 years), are characterized by the use of various digital tools and options, they like to shop through social networks, e.g. Facebook, Instagram, and more. Many of them go to stores before buying it online. They are skilled in digital technologies, which allows them to work with price comparisons and therefore make purchases of their goods at reasonable prices.

The group of the middle generation of 40–50 year olds has a relatively even representation in several shopping centres (Fig. 4). Increased attractiveness is among respondents using OC Galéria in Nitra and OC Aupark and ZOC Max in Žilina. This generation is characterized by an effort to maintain resp. to cope with the younger generation, which is reflected in its imitation of, e.g. maintain fashion, style, be trendy, e.g. in clothing, way of spending free time, etc. This decision-making process provides the basis for the development of the megatrend, known in the literature as the “rejuvenating population”.

Generation 60 and older—seniors, known as baby boomer is characterized by certain shopping habits, e.g. place of purchase, they buy mainly where they are used to. They don’t like changes as they limit them. They are well informed (they know leaflet promotions, discounts, various financial benefits, etc.). They are time-independent, expect comfort and service at high level and skilled workforce. Although they prefer smaller stores, it is not true that seniors do not shop online, the number of shoppers over 65 is even growing year on year (IPSOS, September 2018).

Nitra, Trnava and Žilina are the old centres not only of secondary, but also of higher education, which was also reflected in the structure of respondents according to their education. Education reflects the breadth of knowledge, cognitive abilities and skills used by individuals in everyday life. Most respondents have completed secondary education (Fig. 4). Secondary education ranges from 59% (OC City Aréna) to 81% (OC Mirage). Higher education is represented in the range of 15% (ZOC Max in Žilina)—32% of respondents (OC Aupark). The least represented among the respondents in the individual city centres are respondents with basic education, which is related to the educational structure of the urban population. Most respondents are involved in the work process as employees. In addition, the respondent component also consists of entrepreneurs, who reach a share of 5–20% in shopping centres. Like employees, entrepreneurs find their application in various sectors of the economy, including retail. Nine to thirteen percentage of respondents are in shopping centres of retirement age. Students of secondary and higher education are in the stage of preparation for the work process, whose share in the respondents of shopping centres reaches from one-tenth (OC Mirage) to one (one-third of the respondents in OC City Aréna and ZOC Max).

One of the important factors influencing shopping is the purchasing power of respondents, which is a reflection of their income and the economic situation of their households. The incomes of the population are highly differentiated. In 2018, the level of the average wage was €1013, it was not reached by up to 57% of Slovaks employed full-time. For 18% of Slovaks, the fact was that the gross wage was lower than €500, which in net terms represents an income of approximately €415 [83].

The majority of respondents (67%) manage with incomes of €500.01–1500. This group includes mainly employees and students. Although many students increase their income through various forms of employment, a substantial proportion of them is subsidized by their parents’ funds. This group is slightly dominated (36%) by respondents with incomes of €1000.01–1500. The unemployed, pensioners and also some students have lower incomes than €500. The minimum average wage per employee is €520 (2019). More than €1500.01 in households are mainly held by entrepreneurs and employees in higher positions. Almost one-quarter of respondents belong to this group. Most Slovak seniors are low-income earners. Their income often does not allow them to live at the required level and therefore it often happens that they are subsidized by their working children or continue in the working

Table 7 Share of respondents household income by selected shopping centres in regional cities of Nitra, Trnava and Žilina in %

Household incomes in Eur	Nitra			Trnava			Žilina		
	A	B	C	D	E	F	G	H	I
≤500	9.8	11.8	10.0	2.6	4.2	9.5	11.0	11.1	3.4
500.01–1000	22.0	16.9	9.4	55.1	44.1	52.7	51.4	18.4	31.0
1000.01–1500	40.0	38.8	37.2	28.2	30.9	31.1	27.5	38.4	44.8
1500.01≥	22.0	30.9	33.3	12.8	19.1	4.1	9.2	31.1	20.7
Not stated	6.3	1.7	10.0	1.3	1.7	2.7	0.9	1.1	0.0

Note: A—OC Galéria Nitra, B—OC Mlyny Nitra, C—OC Centro Nitra, D—OC Galéria Trnava, E—OC City Aréna Trnava, F—OC ZOC Max Trnava, G—OC Mirage Žilina, H—OC Aupark Žilina, I—OC ZOC Max Žilina
Source: Questionnaire survey: Department of Geography and Regional Development, Constantine the Philosopher University in Nitra, 2015, processed by the authors

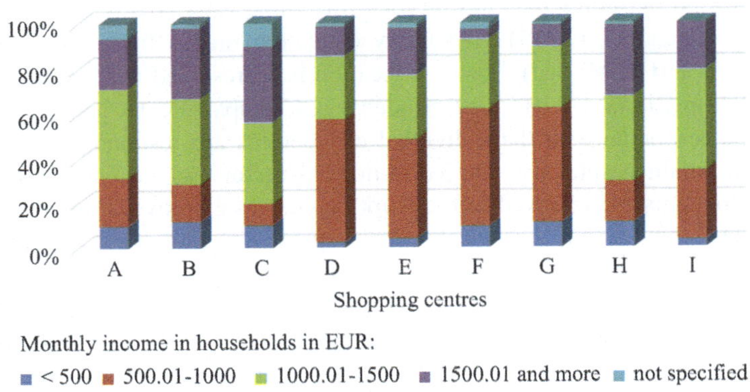

Fig. 5 Share of respondents' household income by selected shopping centres in the regional cities of Nitra, Trnava and Žilina in %. Note: A—OC Galéria Nitra, B—OC Mlyny Nitra, C—OC Centro Nitra, D—OC Galéria Trnava, E—OC City Aréna Trnava, F—OC ZOC Max Trnava, G—OC Mirage Žilina, H—OC Aupark Žilina, I—OC ZOC Max Žilina. (*Source*: Questionnaire survey: Department of Geography and Regional Development, Constantine the Philosopher University in Nitra, 2015, processed by the authors)

process. The picture of respondents according to individual household income is documented in Table 7 and Fig. 5.

The economic situation of a household is influenced by the number of household members as well as the size and income structure of their individual members. The average household size of respondents in shopping centres is almost balanced, ranging from 3.2 to 3.9 household members (Table 8).

As can be seen from Fig. 6 in the most attractive retail centres in Nitra, respondents with a household income of €1000.01–1500 predominate, similarly to the Aupark and Max shopping centres. On the contrary, in the shopping centres of Trnava and OC Mirage in Žilina, the household income of the respondents is lower (€500.01–1000). Respondents with household incomes over €1500.01 like

Table 8 Average household size of respondents by selected shopping centres in regional cities of Nitra, Trnava and Žilina in %

Nitra			Trnava		Žilina			
Galéria	Mlyny	Centro Nitra	Galéria	City Aréna	Max	Mirage	Aupark	Max
3.63	3.87	3.70	3.36	3.47	3.23	3.31	3.74	3.76

Source: Questionnaire survey: KGRR Department of Geography and Regional Development, Constantine the Philosopher University in Nitra, 2014, 2015, processed by the authors

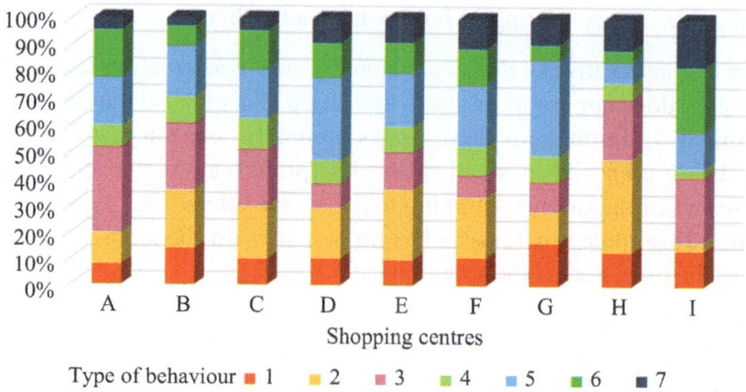

Fig. 6 Types of consumer shopping behaviour in individual shopping centres of regional cities of Nitra, Trnava and Žilina in %. Note: A—OC Galéria Nitra, B—OC Mlyny Nitra, C—OC Centro Nitra, D—OC Galéria Trnava, E—OC City Aréna Trnava, F—OC ZOC Max Trnava, G—OC Mirage Žilina, H—OC Aupark Žilina, I—OC ZOC Max Žilina. Behaviour type: 1—influenceable, 2—demanding, 3—mobile, 4—conservative, 5—sparing, 6—loyal, 7—unpretentious phlegmatic. (*Source*: Questionnaire survey: Department of Geography and Regional Development, Constantine the Philosopher University in Nitra, 2015, processed by the authors)

shopping mainly in OC Centro in Nitra and OC Aupark in Žilina. This group consists mainly of employees in senior positions and entrepreneurs. In Trnava, OC Gallery is the most preferred. Better financially secured consumers are, in essence, more demanding on quality, like to pay extra and are interested in buying more luxurious, fashionable and more expensive goods.

The respondents with lower incomes focus on buying goods where the price equals quality, often buy lower quality goods or buy in markets, especially vegetables and fruits. These respondents are also looking for cheaper shops, e.g. with Chinese goods. In shopping centres, they use a number of financially advantageous offers, e.g. special offers, sales, etc.

The globalization and internationalization of the shopping network have brought a number of new elements and changes to the sales network, which have been reflected in changes in shopping habits. The arrival of foreign retail chains has led to the opening of a number of branded stores, especially in the shopping centres of large cities but also non-commercial services (banks, post offices, travel agencies, etc.). Shopping and entertainment centres have become new "temples of consumption" in which people find not only shops with goods and services, but also spaces for entertainment and distraction (cinemas, games rooms, cafes, bars and restaurants, fitness centres, ice rinks or swimming pools) [82]. Under their influence, consumers' shopping behaviour changes. In addition to the personal and economic attributes analysed above, the respondents' behaviour reflects a number of other objective and subjective factors that determine purchasing, and therefore respondents were asked to classify themselves into one of the subtypes of purchasing. Behaviours, which can be divided into **two basic types:** the first type is traditional behaviour consisting of **four consumer subtypes**:

- **cautious** (conservative)—decides rationally and conservatively, does not trust advertising, does not use a car, has a low share of impulsive purchases
- **economical**—minimizes expenses, buys rationally, uses car just a little
- **loyal**—buys often and in small quantities, prefers a smaller store near home, professes tradition, orients shopping to the social side of life
- **undemanding** (phlegmatic)—he has no demands on the store, he is indifferent to prices, he does not go shopping

The second basic type reflects the modern orientation in consumer behaviour. It consists of **three shopping subtypes:**

- **influenced**—emotional in shopping behaviour, influenced by advertising, buys stock goods, a large number of purchases in 1 day
- **demanding**—has high demands on the quality of goods and shopping comfort
- **mobile** (pragmatist)—optimizes the ratio between the price and value of goods, prefers large stores, regularly uses the car [59]

The consumers' shopping habits change over time and space, and in order for retailers to become ever closer to the consumer, they need constant attention. All types of customers' shopping behaviour are represented in the shopping centres of individual cities (Fig. 6). The modern type of consumer shopping behaviour (53:47) slightly predominates among the respondents, although the situation is different in individual shopping centres. This type of consumer shopping behaviour is prevalent in five of the nine shopping centres in regional cities.

Consumers' shopping habits change over time and space, and so that retail can become more and more. In Nitra, this type includes all three shopping centres (OC Galéria, OC Mlyny, OC Centro), in Trnava only one (OC City Aréna) similar to Žilina (OC Aupark).

There are differences between demographics and types of shopping behaviour. The observed statistical features are qualitative and therefore the χ^2 square contingency test was used to measure the dependence. By testing, we want to find out whether the characters are statistically significant or are only random, i.e. whether there is a relationship between qualitative features. If so, then it is necessary to determine the intensity of this dependence. Two hypotheses were established: null and alternative, which have the form:

H0—there is no statistical dependence between the type of shopping behaviour and the sex of the respondent
H0—there is no statistical dependence between the type of shopping behaviour and the (generation) of the respondent
H0—there is no statistical dependence between the type of shopping behaviour and the education of the respondent
H0—there is no dressing dependence between the type of shopping behaviour and the monthly income of the respondent's household
H1—there is no dressing dependence between the type of shopping behaviour and the sex of the respondent
H1—there is no dressing dependence between the type of shopping behaviour and the (generation) of the respondent
H1—there is no dressing dependence between the type of shopping behaviour and the education of the respondent
H1—there is no dressing dependence between the type of shopping behaviour and the economic activity of the respondent
H1—there is no dressing dependence between the type of shopping behaviour and the number of members in the respondent's household
H1—there is no dressing dependence between the type of shopping behaviour and the monthly income of the respondent's household

As can be seen from Table 9 there is a statistical dependence between individual demographic features and the feature—type of shopping behaviour. From a comparison of the values of the calculated test criterion, which are greater than the critical values, it is clear that the null hypothesis of independence is rejected and that there is a statistical dependence between individual demographics and the type of shopping behaviour. We monitored the intensity of this dependence using the Contingence coefficient and Cramer's *V* Coefficient. The coefficients of dependence take values in the range from 0 to 1. Both coefficients confirmed the dependence between the studied demographic features and the type of shopping behaviour. There is a weak degree of dependence (intensity) between the sign of gender, education, number of members in the household and the type of shopping behaviour and a slight statistical dependence on the sign of generation, economic activity and monthly household income and type of shopping behaviour.

The cities of Slovakia are mostly characterized by modern consumer shopping behaviour, as evidenced by the studied western cities—Nitra, Trnava and Žilina. In the modern type, demanding respondents ($N = 265$, i.e. 21%) have a more significant position, reaching one-fifth (22%) of respondents in regional cities. They focus mainly on the quality, fashion and luxury of mainly branded goods and require shopping comfort, which also allows them higher incomes. The group is dominated by women, the ratio of women to men is 60:40. The educational level in this segment of respondents is very high, up to 52% of respondents have secondary education and 45% have completed university education, which is the highest share among individual subtypes of purchasing behaviour of the entire research file. More than half of the respondents have more than €1000 of this shopping behaviour

Table 9 Influence of demographic characteristics on respondents' consumer behaviour

	Gender	Generation	Education	Economic activity	Number of household members	Households incomes
χ^2 calculated	55,468	205,782	101,840	221,212	97,177	145,090
χ^2 critical	39.63	7.56	5.54	6.00	0.8	8.02
Dependence (given character has an effect)	Has an effect	Has an effect	Has an effect	Has an effect	Has an effect	Has an effect
Phi	0.211	0.407	0.286	0.422	0.280	0.341
Cramer's *V*	0.211	0.182	0.202	0.211	0.114	0.171
Contingency coefficient	0.206	0.377	0.275	0.388	0.269	0.323
Intensity of effect	Low	Low to moderate	Low	Low	Low	Low to moderate

Source: Own research

(of which 43% have more than €1500 and 19% are in the income category of €1000.01–1500). Higher revenues are a reflection of higher positions, especially for employees. Up to 45% of entrepreneurs in the entire research component are included in this group. Demanding respondents are significantly made up of two generations, 20–39 years old and 40–49 years old. At the forefront is the 40–49 year old generation, which tends to match the younger generations and has higher financial incomes. Households of this subtype are characterized by 4–5 members, mainly parents and children.

Eighteen percentage of respondents in Nitra, 24% in Trnava and 25% in Žilina were included in the demanding segment. Demanding respondents in Trnava have a popular shopping experience in OC Galéria, where more than half of them shop (61%). Shopping demanding in ZOC Max (20%) reaches one fifth, similar to shopping in OC City Arena (18%). While in Trnava and Žilina the **demanding type** is primary among respondents, in Nitra it is secondary. Nitra respondents like to use OC Mlyny (39%) and OC Centro (36%) for their purchases. Shopping in OC Gallery OC Mlyny (25%) characterized by lower diversification of non-food assortment as well as a smaller number of branded entities. In the regional city of Žilina, this type is concentrated in OC Aupark, which is looking for more than 83% of demanding customers. In OC Mirage and ZOC Max, only 17% of demanding respondents from the city shop.

The second subtype of modern consumer behaviour is mobile (pragmatic). It is represented by almost one-fifth of respondents ($N = 248$; 20%) of regional cities. When shopping, this subtype of consumers prefers large-scale stores. The pragmatist focuses mainly on goods in which the price equals the value of the goods. More than 60% of respondents are aged 30–50. The ratio of men to women is 47:53. The respondents of the segment are characterized mainly by secondary education. Almost three-quarters of respondents (73%) have completed secondary education and one fifth (24%) have a university degree. These respondents are mainly active as employees and students. Households of pragmatists are characterized by 3–5 members. The monthly income of their household is lower than in the demanding segment but higher than in other subtypes of shopping behaviour. Over €1000, 61% of households have respondents, almost half of whom have monthly incomes above €1500.

Among regional cities in this type, Nitra is dominated (26%), which is twice as large as the group of Trnava respondents (13%). Compared to Žilina pragmatists, they also have a more significant representation (18%). Almost half (44%) of customers of this type prefer shopping at OC Galéria Nitra. Other OCs in Nitra and Žilina have a relatively balanced share of respondents, which ranges from 24.7% to 21.1%. OC in Trnava (graph x) and OC Mirage in Žilina have lower values of the share of mobile customers.

The lowest share of respondents in the modern type belongs to influential customers. This segment consists of more than one-tenth of the research group ($N = 138$ persons; 12%). In shopping behaviour, these customers are emotional, they can quickly be influenced by advertising, they make a larger number of purchases in 1 day, they buy branded goods. This behaviour is especially characteristic for the younger generation of 20–30 year olds. This generation accounts for two-fifths of the total research population (43%). This generation is characterized by emotional purchases. Emotional customers do not buy a specific product or service, not a brand. Rather, they make purchases with caution, buying mainly when something hits their imagination or interests them. They often shop under the influence of ads. The highest share of influential respondents is in Žilina (14%), decimal in Trnava (10%) and also decimal in Nitra (11%) of the total set of respondents. In Žilina, OC Aupark (53%) is the most searched among other influenced shopping centres. In Trnava, half of the respondents of this subtype also shop in OC Galéria (54%). The most popular shopping centres in Nitra are OC Mlyny (43%), OC Centrum (30%) and OC Galéria (27%).

While the modern type is dominated by the demanding subtype in the traditional shopping type, the savings segment has a priority position, which is characteristic especially for low-income groups. Older generations also have an important position in individual subtypes. The older generation tends to the traditional type of shopping, which shows a higher share of thrift, conservatism and loyalty. Forty-seven percentage ($N = 591$) of respondents of the research component were included in the traditional type. Its internal structure is dominated by the savings segment ($N = 243$), which reaches a 20% share in the research group of respondents. Thrifty customers are characterized by minimizing their expenses. They buy rationally, only the specific products they need. They know in advance what they are buying. This type of customer is easily lured at various discounts. They usually do not buy the product at full price. They are looking for a discount on the product they want to buy, they often visit several stores. They are driven by the desire to save money. In this segment, similarly to the impactable one, the generation of 20–29 year olds has a more significant position, reaching a third share (32%) and a fifth share of the senior generation. About one-fifth of this subtype is represented in every city. In Nitra, the share of this segment is almost balanced, reaching a one-third share in individual shopping centres. In Trnava, economical respondents prefer shopping mainly in OC Galéria (49%) and in Žilina in OC Mirage Žilina (66%).

Another segment is the loyal subtype, in which one-fifth (N130; 10.69%) of respondents were included. Respondents of this subtype often shop in small quantities, mostly in stores close to their home. Older respondents have a more significant position in this group. Respondents over the age of 50 achieve up to 46% share in this sample. Due to its loyalty, this segment is valuable for retail stores. Loyal customers can become promoters of the store where they shop or brand among friends, family, on social networks, etc.

The shopping behaviour of a conservative (cautious) customer ($N = 114.9\%$) is characterized by rational decision-making, does not trust advertising and is characterized by a low share of impulsive purchases. The subtype has a more significant representation in the group of 20–30 year olds and 40–50 year olds. These two generations reach up to half of the research component (53%) of the segment. The customer of the subtype undemanding phlegmatic ($N = 97$, 8%) is characterized by the fact that he has no demands on the store, he is indifferent to the prices, he does not go shopping. It occurs in every age category.

Although it can generally be assumed that the younger generations will tend to the modern type, and the older ones to the traditional type, the individual generations are internally so diverse that this translation is ambiguous. E.g. the 20–30 year old generation has a significant position in the modern type as an influential segment (43%), resp. demanding (27%) but also stands out in the traditional as thrifty (32%) and conservative (32%), which belong to the traditional type. The position of the older generation is also changing, and it is becoming much more inclined to modern shopping behaviour. In shopping centres, it is not possible to set aside a one-sided orientation to a certain type of shopping behaviour, but they are characterized by the diversity of customers shopping behaviour.

Although the demographic characteristics of the respondents sample (gender, generation, type of economic activity, number of household members and household income) from testing influence the types of shopping behaviour that results from their testing. However, the degree of dependence between them is low and moderate in intensity.

5 Conclusions

In the context of the retail structures localization development, the urban areas are increasingly confronted with situations of competition between traditional inner cities and spontaneously developing suburban forms. Retail locations with the attribute of a suburban spiral are outlets that, on the one hand, meet the increased demands on retail space, parking and productivity, and on the other hand, citizens, politicians and established ones fear the "death" of a traditional shopping centre. Assessing the impact of planning decisions in this regard is becoming more complex: this can be attributed to a strong individualization of consumer habits as well as increased sensitivity to exposure to emissions from traffic and land consumption.

The larger cities of Slovakia examined in this study, compared to other cities in Central Europe, had a specific development after 1989, surprisingly more similar to the development in large cities of Russia. Based on the research of spatial organization of retail and shopping patterns of behaviour of three regional cities of Slovakia—Nitra, Trnava and Žilina, it is possible to point out identical processes (atomization stage and concentration stage), characteristic of cities in Central Europe, but also specific, typical for Slovak cities (consolidation stage and saturation stage). Already in 2008, the initial manifestations of the qualitative next stage were identified, which [51] described it as demasification.

Based on the regularities of the development of the spatial organization of retail measured by the area parameter (PAFS) in the cities of Nitra, Trnava and Žilina, the transformation of the monocentric model of the MO network to a polycentric in the concentration stage was identified. While in the last decade of the twentieth century, retail entities were specifically concentrated in the traditional nuclear-trade zone, due to the processes of globalization and internationalization, new centres of retail services were created on the outskirts of cities. Three cores were identified in the regional city of Nitra and two in Trnava and Žilina.

As in other cities in Slovakia, there are trends in studied cities, such as transfer of construction from city centres (inner city) to their outskirts or to suburbs (outer city) located mainly on green areas, sharpening the competitive struggle for the customer, modernization and remodelling of shopping centres, e.g. reducing the size of their sales area in favour of leisure services, etc.

The location of shopping centres close to the city's cores can represent a certain opportunity for the city to develop, but also a certain risk. On the positive side, despite the construction of shopping centres in city centres with a historical tradition such as Trnava and Žilina, the developed MO network has been preserved and traditional pedestrian shopping streets have not been depopulated. In accordance with Hillier's theory of centrality (1999), they managed to preserve a "living centre", which cannot be said in case of ancient Nitra, where the pedestrian zone was significantly subdued. The location and transfer of retail activities from the shopping centre in the city centre to the newly built OC Mlyny (which was created on the site of a historic industrial brownfield, after which only the name of the shopping centre has been preserved), led to its secondhandisation, low-cost goods—the centre is not able to maintain the full range, it is depopulated during the day. Retailers seeing higher profits, competitive advantages based on modernity and concentration of activities in one place as well as the possibility of using large parking spaces, they moved their stores to the shopping centre. In contrast, the construction of the Arena City shopping centre in Trnava increased the attractiveness of the centre. Positive interaction of new allocated retail centres in relation to small sales entities—diversification of retail forms helps to a lively city centre.

Shopping centres have an increasingly strong position in Slovak retail, they participate to a large extent in retail turnover and their number is constantly increasing [45]. They influence the customers shopping behaviour. The way and type of consumer shopping behaviour is also changing from traditional to modern. The goal of shopping is also changing. Customers buy not only out of necessity, but also for fun, pleasure. The goal and method of shopping is also changing. Customers buy not only out of necessity, but also for fun, pleasure. Shopping is influenced by a number of factors and conditions, the degree and influence of action is different. They are comprehensively reflected in the shopping behaviour of customers. Due to the number of factors that enter the purchasing process, a huge number of typologies exists. Which can basically be synthesized into traditional and modern shopping behaviour. Modern forms of shopping behaviour, will suppress even replace traditional forms of shopping by its development.

The currently spreading development of technologies in MO as a reflection of the increased electronic "literacy" of the population today, which also leads to negatives, such as to the "dehumanization" of MO stores by reducing the number of employees in shops. E-shops, e-comers as a fashion trend compete with stores that must constantly improve the quality of sales services, adapt existing offers to new trends and remodel store formats to take into account the changing demands of consumers. We agree with the statement of [5] that the stratification properties of shopping—shopping types, the dominance of certain types of facilities and basic macro-regions in the spatial organization of the tertiary sector consists of several stages and is closely linked. Retail thus appears as a variable, dynamic subsystem of the cultural landscape, which is sensitive to changes in the social and economic situation, and is a suitable indicator of the human society economic development [27]. These phenomena can manifest themselves in the longer term in the spatial and functional structure of settlements. This should be taken into account in advance in the production of spatial plans, territorial systems of ecological stability and various development concepts.

For various reasons, these dynamic areas of retail restructuring pose the greatest challenge to achieving planning and urban policy and new approaches to public control over retail spatial processes.

Acknowledgement This article was prepared as part of the VEGA No. 1/0799/14 project: "Geographical Aspects of the Retail Network of Large Cities in New Market Conditions", APVV-18-0185 Land-use changes of Slovak cultural landscape and prediction of its further development and within scheme H2020 project: "Linking Research and Innovation for Gender Equality", Nr. 873134.

References

1. I. Tosics, Determinants and consequences of spatial restructuring in postsocialist cities, in *ENHR Conference*, 2–6 Jul (University of Cambridge, University of Illinois, 18–19 Jun 2004)
2. T. Nejedlý, *Obchody budúcnosti tvoria aj Slováci. Trend: týždenník o ekonomike a podnikaní*, vol. 27(17) (News and Media Holding, Bratislava, 2017), pp. 43–45
3. J. Meľuch, T. Jurášková, *Technológie menia maloobchod. Trend: týždenník o ekonomike a podnikaní*, vol. 27(17) (News and Media Holding, Bratislava, 2017), pp. 3–44
4. M. Trembošová, Vybrané aspekty transformácie maloobchodu v meste Nitra v rokoch 1992–2008. Geografický časopis **62**(1), 49–73 (2010)
5. M. Trembošová, *Geografické aspekty maloobchodnej siete mesta Nitra* (Prírodovedec UKF, Nitra, 2012)
6. Ľ. Nagyová. Zmeny na slovenskom trhu potravín v dôsledku vstupu zahraničných obchodných sietí, in *Zborník z medzinárodnej konferencie Ekonomika Firiem 2000*, (Podnikovohospodárska fakulta Ekonomická univerzita, Košice, 2000), pp. 451–456
7. M. Vicen, Marketingová filozofia a stratégia, in *Marketing*, ed. by A. Kretter et al., (SPU, Nitra, 2004)
8. J. Kunc, F. Križan, Changing European retail landscapes: New trends and challenges. Moravian Geogr. Rep. **26**(3), 150–159 (2018)
9. P.W. Daniels, *Service Industries: A Geographical Approach* (Methuen & Co. Ltd., London, 1985)
10. L. Salvaneschi, *Location, Location, Location: How to Select the Best Site for Your Business* (Oasis Press/PSI Research, Grants Pass, 1996)
11. K. Jones, J. Simmons, *Location, Location, Location. Analyzing the Retail Environment*, 2nd edn. (Nelson Canada, Scarborough, 1993)
12. E.J. Miller, M.E. O'Kelly, Estimating shopping destination choice models from travel diary data. Prof. Geogr. **35**(4), 440–449 (1993)
13. B. Hillier, Centrality as a process: Accounting for attraction inequalities in deformed grids. Urban Des. Int. **4**(3–4), 107–127 (1999)

14. R.L. Church, A. Murray, *Location Covering Models. History, Applications and Advancements* (Springer International Publishing, Cham, 2018)
15. E. Nagy, Winners and losers in the transformation of city centre retailing in east Central Europe. Eur. Urban Region. Stud. **8**(4), 340–348 (2001)
16. C. Guy, Classifications of retail stores and shopping centres: Some methodological issues. GeoJournal **45**(4), 255–264 (1998)
17. F. Križan, K. Bilková, P. Barlík, P. Kita, M. Šveda, Old and new retail environment in a post-communist city: Case study from the old town in Bratislava, Slovakia. Ekonomický časopis **67**(8), 879–898 (2019)
18. J. Lawrence, M. Kuby, The location types of US retailers. Int. J. Appl. Geospat. Res. **7**(4), 1–22 (2016)
19. E. Nagy, Fall and revival of city centre retailing, in *Planning an Urban Function in Leicester, UK* (1999)
20. S. Porta, V. Latora, F. Wang, S. Rueda, E. Strano, S. Scellato, A. Cardillo, E. Belli, F. Cardenas, B. Cormenzana, L. Latora, Street centrality and the location of economic activities in Barcelona. Urban Stud. **49**(7), 1471–1488 (2012)
21. O. Rotem-Mindali, Retail fragmentation vs. urban livability: Applying ecological methods in urban geography research. Appl. Geogr. **35**(1–2), 292–299 (2012)
22. N. Kshetri, Entrepreneurship in post-socialist economies: A typology and institutional contexts for market entrepreneurship. J. Int. Entrep. **7**(3), 236–259 (2009)
23. Z. Szczyrba, *Maloobchod v ČR po roce 1989*, 1st edn. (Univerzita Palackého, Olomouc, 2005)
24. M. Trembošová, A. Dubcová, Vývoj maloobchodu v Nitre v nových trhových podmienkach 2013. Acta Geographica Universitatis Comenianae **57**(2), 213–230 (2013)
25. W. Wilk, The services sector in the new socio-economic reality, in *Poland in the Geographical Centre of Europe. Political, Social and Economic Consequences*, ed. by M. Czerny, (Nova Science Publishers, Hauppauge, New York, 2006), pp. 165–184
26. A. Aubert, J. Csapó, Transition of retail supply of the small cities of Baranya country, in *Neue Einzelhandelstrukturen am Rande von Kleinstädten, Tagung Forschungssechseck Hartberg*, Graz (2004)
27. K.E. Aksenov, Evolution of the types of shopping and spatial organization of retail trade in the post-Soviet metropolis. Reg. Res. Russ. **6**(4), 375–386 (2016)
28. K.E. Aksenov, D. Krupickaitė, K. Morachevskaja, A. Zinovyev, Retail sprawl in Soviet urban residential communities (microrayons): Cases of Saint-Petersburg and Vilnius. Moravian Geogr. Rep. **26**(3), 210–219 (2018)
29. D. Pilat, Regulation and performance in the distribution sector. Working Paper No. 180 (OECD Economic Dpt., Paris, 1997)
30. Ennis. Ennis and Environs Development Plan 2008—2014. Retail Strategy for the Mid West Region, vol. 2(5) (2008)
31. C. Teller, S. Wood, A. Floh, Adaptive resilience and the competition between retail and service agglomeration formats: An international perspective. J. Mark. Manag. **32**(17–18), 1537–1561 (2016)
32. M. Bertrand, F. Kramarz, Does entry regulation hinder job creation? Evidence from the French retail industry. Q. J. Econ. **117**(4), 1369–1413 (2002)
33. J. Haskel, R. Sadun, Regulation and UK retailing productivity: evidence from micro data. CEPR Discussion Paper 7140 (2009)
34. F. Schivardi, E. Viviano, Entry barriers in retail trade. Econ. J. **121**(551), 145–170 (2011)
35. K. Baar, *Legislative Tools for Preserving Town Centres and Halting the Spread of Hypermarkets and Malls Outside of Cities. Draft for Review: March 2002* (Institute for Transport and Development Policy, New York, 2002)
36. L. Klapper, L. Laeven, R. Rajan, Entry regulation as a barrier to entrepreneurship. J. Financ. Econ. **82**(3), 591–602 (2006)
37. T. Borén, M. Gentile, Metropolitan processes in post-communist states: An introduction. Geografiska Annaler: Séria B, Human Geographie **89**(2), 95–110 (2007)

38. D. Pojani, Urban and suburban retail development in Albania's capital after socialism. Land Use Policy **28**(4), 836–845 (2011)
39. M. Cvetinovič, D. Veselinovič, Instrumental participation in Serbia: Online platform for the dialogue about public spaces, their availability and public usage, in *Technologies for Sustainable Development*, (Springer, Cham, 2014), pp. 159–170
40. S. Tsenkova, Planning trajectories in post-socialist cities: Patterns of divergence and change. Urban Res. Pract. **7**(3), 278–301 (2014)
41. P. Hlaváček, P. Raška, M. Balej, Regeneration projects in central and eastern European post-communist cities: Current trends and community needs. Habitat Int. **56**, 31–41 (2016)
42. M. Maksić, Institutional obstacles in large-scale retail developments in the post-socialist period—A case study of Niš, Serbia. Cities **55**, 113–121 (2016)
43. I. Košičiarová, Ľ. Nagyová, Z. Kádeková, M. Holienčinová, J. Rybanská, *Manažment obchodnej spoločnosti*, 1st edn. (Key Publishing, Ostrava, 2017)
44. Ľ. Kubicová, Z. Kádeková, *Strategic Marketing*, 1st edn. (Slovak University of Agriculture, Nitra, 2017)
45. J. Mitríková, *Geografia maloobchodu (vybrané kapitoly)*, 1st edn. (Bookman, Prešov, 2017)
46. D.M. Scott, S.Y. He, Modeling constrained destination choice for shopping: A GIS-based, time-geographic approach. J. Transp. Geogr. **23**, 60–71 (2012)
47. M. Franz, A. Appel, M. Hassler. Short waves of supermarket diffusion in Turkey. Moravian Geogr. Rep. **21**(4), 50–63 (2013)
48. J. Hultman, U.C. Johansson, A. Wispeler, L. Wolf, Exploring store format development and its influence on store image and store clientele—The case of IKEA's development of an inner-city store format. Int. Rev. Retail Distrib. Consum. Res. **27**(3), 227–240 (2017)
49. D. Grewal, A.L. Roggeveen, J. Nordfält, The future of retailing. J. Retail. **93**(1), 1–6 (2017)
50. E.T. Bradlow, G. Manish, K. Praveen, S. Voleti, The role of big data and predictive analytics in retailing. J. Retail. **93**(1), 76–96 (2017)
51. M. Trembošová, P. Tremboš, Etapy vývoja maloobchodnej siete mesta Nitra v období rokov 1992–2008. Acta Geographica Universitatis Comenianae **53**(1), 123–138 (2009)
52. S. Kabadayi, B.A. Pakson, Segmentation of Turkish consumers based on their motives to visit shopping centres. Int. Rev. Retail Distrib. Consum. Res. **26**(4), 456–476 (2016)
53. S. Alavi, S. Rezaei, N. Valaei, W.I. Wan Khairuzzaman, Examining shopping mall consumer decision-making styles, satisfaction and purchase intention. Int. Rev. Retail Distrib. Consum. Res. **26**(3), 272–303 (2015)
54. M. Puelles, M. Diaz-Bustamante, S. Carcelén, Are consumers more rational and informed purchasers during recession periods? The role of private labels and retailing strategies. Int. Rev. Retail Distrib. Consum. Res. **26**(4), 396–417 (2016)
55. G. Cristini, F. Laurini, Growth factors of store brands in different store formats in Italy. Int. Rev. Retail Distrib. Consum. Res. **27**(2), 109–125 (2016)
56. J. Jantarat, R. Sharon, The moderating effects of in-store marketing on the relationships between shopping motivations and loyalty intentions. Int. Rev. Retail Distrib. Consum. Res. **26**(5), 566–583 (2016)
57. G. Warnaby, D. Bennison, B.J. Davies, Marketing town centres: Retailing and town centre management. Local Econ. **20**(2), 183–184 (2005)
58. J. Mitríková, A. Šenková, S. Antoliková, Application of the huff model of shopping probability in the selected stores in Prešov (Prešov, the Slovak Republic). Geographica Pannonica **19**(3), 110–121 (2015)
59. V. Stárová, *Segmentace nákupního chování—trend i příležitost. Výsledky shopper typology & media behaviour* (Incoma Research a Gfk, Praha, 2003)
60. A. Bertaud, R. Buckley, *Cracow in the Twenty First Century: Prince of Merchants? A Cityś Structure Under the Conflicting Influences of Land Markets, Zoning Regulations and Socialist Pasr* (Mimeo, Budapest, 1997)
61. M. Trembošová, A. Dubcová, M. Civáň, *Geografické aspekty maloobchodnej siete mesta Trnava a nákupné správanie jej obyvateľstva*, 1st edn. (Prírodovedec UKF, Nitra, 2016)

62. M. Trembošová, A. Dubcová, H. Kramáreková, Ľ. Nagyova, *Nákupné správanie obyvateľstva mesta Nitry—vývoj a zmeny*, 1st edn. (SPU, Nitra, 2016)
63. P. Beluszky, The Hungarian urban network an the end of the second millenium. Discusion papers. Centre for regional Studies of the Hungarian Academy of Sciences 44 (1999), p. 39
64. W. Wilk, The services sector in the new socio-economic reality, in *Poland in the Geographical Centre of Europe. Political, social and economic consequences*, ed. by M. Czerny, (Hauppauge, Nova Science Publishers, New York, 2006), pp. 165–184
65. J. Mitríková, *Geografické aspekty transformácie maloobchodu a nákupného správania sa na Slovensku (prípadové štúdie z miest Prešova Košice)*, 1st edn. (Prešovská univerzita, Prešov, 2008)
66. G. Weclawowicz, *Miasto polskie v transformacji—ksztaltowanie sie miasta postsocialistycznego* (Zróznicovanie prszestrzenne struktur spolocznych w duzych miastach, Lodž, 1990)
67. B. Pokorska, E. Maleczcyk, *Koncentracija i integracja w handlu wewnetrznym* (PWN, Warszawa, 2002)
68. A. Toffler, H. Tofflerová, *Utváranie novej civilizácie. Politika Tretej vlny* (Open Windows, Bratislava, 1995)
69. M. Trembošová, A. Dubcová, M. Štubňová, The specifics of retail network's spatial structure in the city of Žilina, 2019. Geographia Cassoviensis **13**(2), 228–245 (2019)
70. J. Mitríková, A. Šenková, S. Antoliková, V. Parová, The analysis of shopping behaviour in the context of spending leisure time activities of consumers in the chosen shopping centres in Vienna (Austria). Економічний часопис **21**(161), 71–74 (2016)
71. F. Križan, M. Zeman, K. Bilková, P. Kita, P. Barlík, Cezhraničné nákupné správanie spotrebiteľov zo Slovenska: Prípadová štúdia z Hainburg an der Donau (Rakúsko). Geographia Cassoviensis **11**(2), 124–136 (2017)
72. M. Solomon, G. Bamossy, S. Askegard, M.K. Hogg, *Consumer Behaviour: A European Perspective*, 3rd edn. (Prentice Hall, London, 2006)
73. J.Z. Sojka, J.L. Giese, Using individual differences to detect customer shopping behavior. Int. Rev. Retail Distrib. Consum. Res. **13**(4), 337–353 (2003)
74. M. Korenkova, M. Maroš, M. Levicky, M. Fila, Consumer perception of modern and traditional forms of advertising. Sustainability **12**(9996), 1–25 (2020)
75. K.B. Monroe, J.P. Guiltinan, Identifying and analyzing consumer shopping strategies. Adv. Consum. Res. **7**, 745–748 (1980)
76. P.H. Bloch, G.D. Buce, The leisure experience and consumer products: an investigation of underlying satisfactions. J. Leisure Res. **16**(1), 74–88 (1984)
77. J. Sit, Understanding youth motivations of entertainment consumption at shopping centres: Implications for shopping centre marketing, in *Conference Proceedings Adelaide*, 1–3 (University of Queensland, Brisbane, 2003), pp. 300–304
78. J. Spilková, *Geografie maloobchodu a spotřeby*, 1st edn. (Karolinum, Praha, 2012)
79. D. Klepochová, Z. Francová, *Imidž nákupného centra a jeho vplyv na spotrebiteľské správanie* (Vedecké state Obchodnej fakulty, Bratislava, 2011)
80. F. Križan, K. Bilková, P. Kita, J. Kunc, P. Barlík, Nákupné centrá v Bratislave a atribúty ovplyvňujúce preferencie spotrebiteľov. Geografický časopis **67**(4), 341–357 (2015)
81. D. Lesáková, Tendencie spotrebiteľského správania a spotreby v domácnostiach dôchodcov na Slovensku. Ekonomický časopis **62**(8), 823–841 (2014)
82. M. Zeman, Sociokultúrny profil návštevníkov nákupných centier na Slovensku: zistenia z empirického prieskumu, in *Maloobchod a špecifiká časovo-priestorového správania spotrebiteľov*, ed. by F. Križan, K. Bilková, P. Barlík, (Univerzita Komenského, Bratislava, 2017), pp. 53–70
83. Retailmagazin, http//www.retailmagaznin.skspotrebitel/trendy/1016-priemery-zarobok-umoznuje-nakupit-viac-nez-v-roku-1989. Last accessed 20 Nov 2019

Developing Car Accident Detecting System Based on Machine Learning Algorithms Applied to Video Recordings Data

Anton Naidenov and Anton Sysoev

1 Relevance of Developing Car Accident Detecting Systems

1.1 Meaning and Potential of the Problem for Russia

The important reason to develop car accident detecting system is not only applying it to the transportation monitoring system but is also using artificial intelligent approaches in the sphere of unmanned vehicle. This idea is the leading one in the road maps of Russian National Technological Initiative (NTI). NTI is the program to support perspective developing Russia economics sectors, which can be the basis of world economy in next 20 years [1]. According to the long-time analysis and based on studying the sphere of economy modernization and innovative development, the Russian government has approved the following road maps of the related world economy markets: EnergyNet, HealthNet, NeiroNet, MariNet, AutoNet, AeroNet, and the cross-market trend TechNet.

Road maps are regulatory instruments and plans of action connected with creating new products, business models, and performing lots of tasks. For example the regulatory legal framework, intelligent transport, and unmanned driving system development providing. The reported study refers to AutoNet market, the main directions of which are unmanned vehicles and intelligence transportation systems. All its aims and ways of development are presented in normative road maps of AutoNet market [2] containing specific events connecting tasks, time of realization, clients, and required resources. This justification determines that the problem under consideration is an actual one and is supported by the federal government.

A. Naidenov (✉) · A. Sysoev
Department of Applied Mathematics, Lipetsk State Technical University, Lipetsk, Russia
e-mail: anton_syssoyev@mail.ru

D. Cagáňová et al. (eds.), *Advances in Industrial Internet of Things, Engineering and Management*, EAI/Springer Innovations in Communication and Computing,
https://doi.org/10.1007/978-3-030-69705-1_4

1.2 Existing Car Accident Detecting Systems

All published projects in this area are based on constructing the certain recurrent neural network (RNN), which is a type of neural network model including cycle inside itself, functioning like an inner memory cell [3]. Such network can use previous state to make predictions more precisely.

1.2.1 Anticipating Accidents in Dashcam Videos

This project was released in 2016 by students of National Tsing Hua University in Taiwan and University of Washington in the USA. They proposed a dynamic-spatial-attention (DSA) RNN to anticipate accidents in dashcam videos [4].

As it is presented on Fig. 1 the video was divided into frames and after these obtained frames pass throw embedding layer. It is a simple matrix addition that transforms object into an appropriate attachments. In other words, it performs object as tensors consisting of numbers. The next step is to apply the attention layer. The attention mechanism is a way to select part of input data (they could be either text fragment or group of pixels) for detailed processing. After the attention process weights of each frame are multiplied with embedding layer. In parallel to this process the specific frame is passing the embedding layer. Further it is concatenated with the global representation. The final step is to transfer existing object to the LSTM (long-short-term-memory) layer. Finally, authors present the neural network taking into account its previous conditions. This structure demonstrates results 80% on a train dataset and 56% on test dataset.

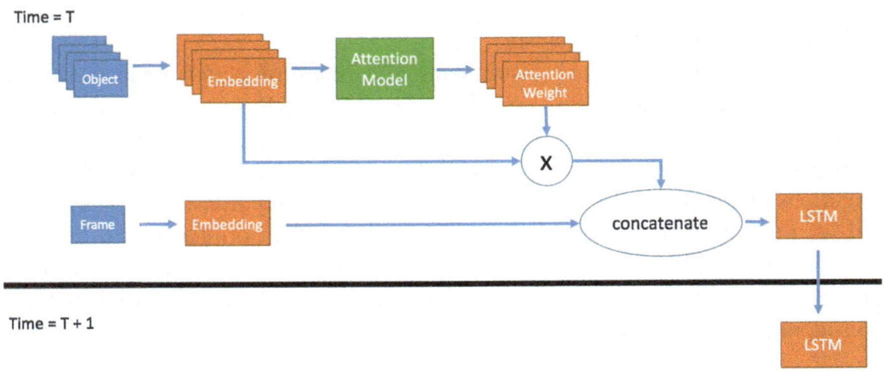

Fig. 1 Anticipating accidents in dashcam videos. Network scheme

1.2.2 CrashCatcher

This project was released by Ph.D. Rachel Wagner-Kaiser from University of Florida in 2017 [5]. It is also based on applying an RNN structure. A hierarchical recurrent neural network algorithm is used to tackle the complex problem of classifying video footage. Firstly video is divided into frames and downscaled in five times. Additionally, it is decoded into shades of gray. It is necessary to accelerate training speed not spending time on color decoding.

Working Principles

- Based on a hierarchical RNN [6].
- The algorithm uses two layers of LSTM neural networks.
- The first neural network is a recurrent network analyzing the time-dependent sequence of images within each video.
- On the second step the neural network takes the encoding of the first model and builds the second neural network that reflects which videos contain accidents and which do not.

Overall accuracy is over 80% for the training, validation, and test sets. The performance and accuracy of the HRNN demonstrates that it generalizes fairly well to videos it has not seen before. It was accurately identified the vast majority of the positive examples in the test set as crashes. In addition, there is a possibility to import videos from YouTube. Receiver operating characteristic is presented on Fig. 2.

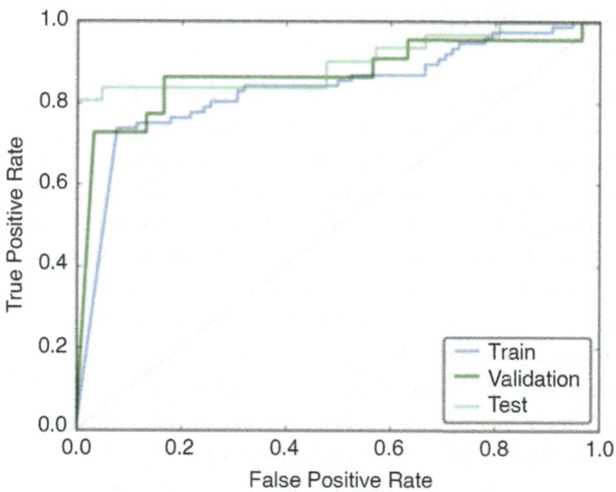

Fig. 2 Crash Catcher. Receiver operating characteristic

2 Building Neural Network

2.1 Design of the System. Used Tools

This study is based on developing the script using Python programming language. Actually it is convolution neural network (CNN). This type of models is becoming more and more popular because of high demonstrated recognition results in the sphere of computer vision. The name "convolution" comes from the operation of convolution, which is a weighted sum of values by pixels intensity [7]. The developed script contains functions for correct working: training, testing, and preprocessing of raw data. Numerical experiments were conducted using additional software components and needed libraries.

Software

- Python 3.7.3—Python language interpreter
- JetBrains PyCharm 2019.2—IDE
- NVIDIA Cuda—is a parallel computing platform and application programming interface (API).

Python Libraries

- TensorFlow—machine learning library
- Keras—machine learning library, add-on for TensorFlow.

2.2 Network Architecture

Basically it is possible to select two main parts of network: convolution basis and fully connected layer. The convolution basis is presented by one of the most applicable architecture VGG16, which is a pretrained at 1.4 billion samples convolution neural network. This structure can recognize more than 1000 classes of objects on different images. Such architecture was developed in 2014, but there are lots of other implementations such as VGG, Inseption, ResNet, Xception, and Inseption-ResNet[8] It is known, that VGG16 is not the fastest one, but it provides a high efficiency and it was the reason that has justified the choice.

Table 1 presents a summary about every stack of layers.

It is clear, that actually VGG16 is a combination of conv2D and MaxPooling2d layers. Conv2d layer is responsible for convolution and MaxPooling2d provides sub-sampling. Gradually image is transformed from the shape (N, 150, 150, 3) into the shape (N, 4, 4, 12), where N is the number of samples. This model has 14,714,688 parameters of learning. Large number of parameters demonstrate complexity of model. VGG16 visualization is presented on Fig. 3.

Table 1 VGG16 summary

Layer (Type)	Output shape	Number of param
input_l	(None, 150, 150,3)	0
block1_conv1 (Conv2D)	(None, 150, 150, 64)	1792
block1_conv2 (Conv2D)	(None, 150, 150, 64)	36,928
block1_pool (MaxPooling2D)	(None, 75, 75, 64)	0
block2_conv1 (Conv2D)	(None, 75, 75, 128)	73,856
block2_conv2 (Conv2D)	(None, 75, 75, 128)	147,584
block2_pool (MaxPooling2D)	(None, 37, 37, 128)	0
block3_conv1 (Conv2D)	(None, 37, 37, 256)	295,168
block3_conv2 (Conv2D)	(None, 37, 37, 256)	590,080
block3_conv3 (Conv2D)	(None, 37, 37, 256)	590,080
block3_pool (MaxPooling2D)	(None, 18, 18, 256)	0
block4_conv1 (Conv2D)	(None, 18, 18, 512)	1,180,160
block4_conv2 (Conv2D)	(None, 18, 18, 512)	2,359,808
block4_conv3 (Conv2D)	(None, 18, 18, 512)	2,359,808
block4_pool (MaxPooling2D)	(None, 9, 9, 512)	0
block5_conv1 (Conv2D)	(None, 9, 9, 512)	2,359,808
block5_conv2 (Conv2D)	(None, 9, 9, 512)	2,359,808
block5_conv3 (Conv2D)	(None, 9, 9, 512)	2,359,808
block5_pool (MaxPooling2D)	(None, 4, 4, 512)	0

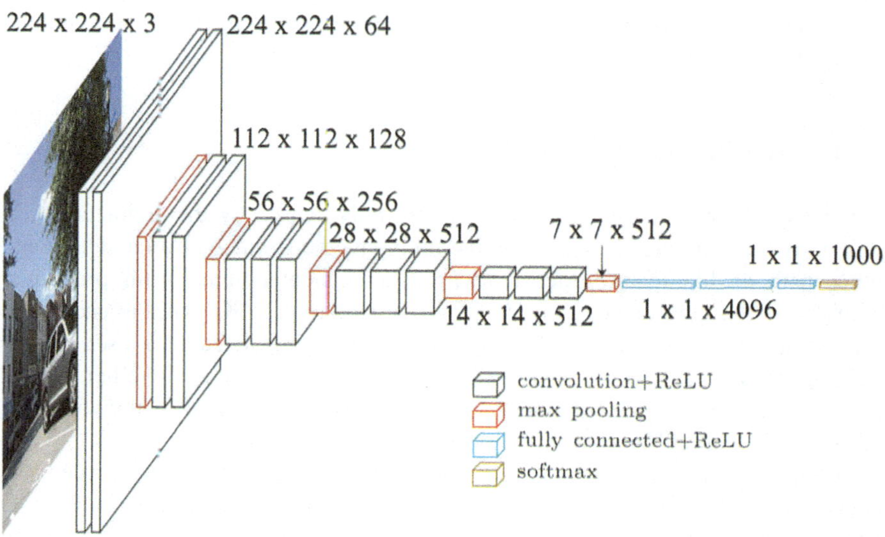

Fig. 3 VGG16 visualization

It is possible to put VGG16 model completely as a layer [9]. Keras has a rich abilities and allows to do this operation easily. The final model architecture contains 16,812,353 and is presented in Table 2.

Table 2 Final model

Layer (Type)	Output shape	# param
vgg16 (Model)	(None, 4, 4, 512)	14,714,688
flatten_1 (Flatten)	(None, 8192)	0
dense_1 (Dense)	(None, 256)	2,097,408
dense_2 (Dense)	(None, 1)	257

The flatten layer in this case is a layer of alignment providing the connection between another layers. Actually it removes all dimensions except one. Two stacks of dense layers performs role of fully connected classifier.

The metadata of the model:

- 100 epochs
- 35 stacks of layers

According to source [10] such number of epochs has demonstrated a high quality and is set as a default.

2.3 *Building Methods*

The main aspect during constructing the neural network model is final accuracy. The analysis of methods answers to the question, why chosen architecture has given the necessary accuracy.

2.3.1 Using Pre-trained Neural Network

Obviously, the reached good accuracy is not only caused by the architecture. In this study the main idea is to use the fine-tuning method, which also includes some other methods, like feature extension with data augmentation and using pretrained network. In this case one can use VGG16 as a good tool. VGG16 has already been trained at ImageNET dataset. Scaling dataset and generalized data used during training of VGG16 allows using spatial hierarchy of features. In other words, pretrained network is the basis for another network, even if another network uses classes, which are different from original data.

Actually it uses a new classifier for learning. The difference between the convolution basis and the classifier is in its current information. The “old” basis keeps general features, which can be used with another data, but it is impossible to use the “old” classifier because of containing information about old classes of images due uniqueness for each task. Also VGG16 does not contain information about car accidents. That is why for obtaining better result during retraining procedure it is a good idea to fine-tune the last block of the convolution basis, but it is necessary to freeze all convolution basis and train a new classifier before.

2.3.2 Extension Data with Augmentation

After defining the convolution basis it was accepted to use extension data with augmentation. In this case every image passes through convolution basis every time, unlike without extension. This process can help to increase calculations significantly, consequently the accuracy also increases. Being a method to decrease the overfitting, Data extension has a lot of reasons, but one of them is the lack of data. The generator that creates new images from original images by squeeze, rotation, and offset became a solution, and consequently the network never sees the same particular image.

2.3.3 Fine-Tuning

Above were presented main steps which are necessary for fine-tuning: adding new network over pretrained, freezing of convolution basis, training new classifier. The final step is to unfreeze the last block of the convolution basis and to retrain it with the classifier. Information about representations in this block and greater number of features help to increase the accuracy. Actually the fine-tuning adjusts features for a new task.

3 Car Accident Detecting System Using Data from Video Recording

3.1 Dataset

For holding this study it was used the VsLab dataset from Anticipation accidents in dashcam videos research. It was released at Asian Conference of Computer Vision in 2016. Dataset contains records with and without car crash moments. It consists of 1730 videos, divided randomly into train and test parts. Train part includes 1284 videos, 455 of them with positive and 829 negative samples. Similarly test part includes 466 records with 165 positive and 301 samples. In addition, test part was divided into test and validate subparts including records from local news web-portals [11]. Examples of video in dataset is presented on Fig. 4 and on Fig. 5. The difference between these pictures is the origin of data. In the first case it is a part of VsLab research dataset that contains car accidents from Taiwan. In the second case it is a part of our own dataset with car crash events in Russia that was taken from local internet resources. The first one is used to train network and other one to look how network will work in action.

Fig. 4 An example of frame from the train dataset sample

Fig. 5 An example of frame from the test dataset sample

3.2 Data Pre-processing

Before loading video into the neural network model, it is necessary to solve what kind of data should be used. It is impossible to load raw video. There are a few ways to handle it. The first way is to divide each video into 100 frames and load them into the network, but weight of dataset do not decrease vastly. A python script was used to separate original data to frames in form of pictures. Then about a few dozens of gigabytes remain on HDD. Next step is to choose one frame from each video. For positive samples it must be the moment of accident and random frame from negative samples. The next question is how to find the right frame in a large dataset? The conducted analysis shows that the car accident moment situated in 70–76 frames within the studied dataset. This number was empirically chosen during frame-divided data analysis. It was decided to take the frame with the number 74. The obtained frames were downscaled to size 150×150 pixels, which helps to reduce the size of dataset and to accelerate speed of the neural network model.

3.3 Results of Numerical Experiments

The study was conducted on the PC with a system requirements:

- CPU: Intet(R) Xeon(R) Silver 4116 CPU @ 2.10 GHz. Number of cores: 24;
- GPU: NVIDIA Tesla P4. NVRAM size: 8 Gb;
- RAM size: 32 Gb;
- HDD size: 1 Tb;
- OS: Windows 10 64-bit.

The time spent to perform of the neural network is about 2 h 56 min. In comparison with previous projects it is a slow network. ACCV for example performs training process less than 1 h. Such long time could be explained by the RMSProp loss optimization function. This exact function was chosen because of working slowly smoothly changing features. Another optimizator could damage them. As for result of training and testing the proposed model it is possible to mention that network has recognized all of 232 test dataset images. Results that were obtained by applying the proposed neural network model are 97% on train dataset, 59% on validation dataset, and 62% on test dataset. Loss value is 0.0828 on train dataset and 2.2582 on validate dataset. ACCV shows results 80% on a train dataset and 56% on test dataset. CrashCatcher has overall accuracy is over 80% for the training, validation, and test sets. According to comparison above it is available to rate proposed network results as acceptable. It is necessary to mark that it is not the best result. If retrain network, it is possible to increase result on a few percent or even more. On Figs. 6 and 7 are shown plots of accuracy and loss on train and test dataset, respectively.

4 Conclusion and Outlook

It is possible to conclude that a car accident recognizing decision support system using machine learning methods and based on video capture recordings was developed. In addition to VsLab dataset own car accidents dataset were used with samples from Russia. Well-known architecture VGG16 learned to recognize

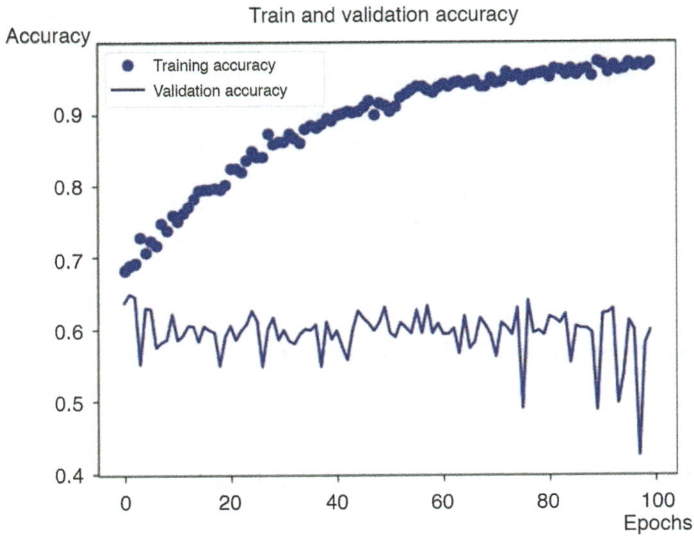

Fig. 6 Recognition accuracy plot

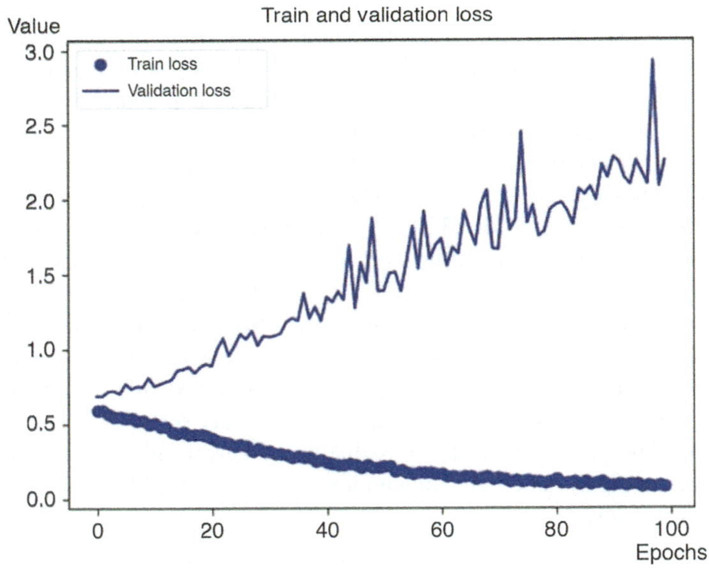

Fig. 7 Recognition loss plot

new class of objects: car accidents that were never used before. Accuracy of the neural network is 97% on train dataset and 62% on test dataset. There are a lot of perspectives of applying such system, for example, as a traffic monitoring system. Shown model is expected to be improved for delivering better results. This study could be promising at fast reaction on accidents and building routes for emergency services. It is the most likely case of applying of this project.

Acknowledgments The reported study is supported by the Russian Science Foundation within the project 18-71-10034.

References

1. National Technological Initiative (NTI). [In Russian: Nacionalnaya Tehnologicheskaya Iniciativa]. http://www.nti2035.ru. Last accessed 13 June 2019
2. Normative road maps of Autonet. [In Russian: Normativnye dorozhnye karty Avtonet]. http://www.nti2035.ru/documents/docs/NDK_autonet.pdf. Last accessed 13 June 2019
3. F. Chollet, *Deep Learning with Python* (Manning, New York, 2017)
4. F.-H. Chan, Y.-T. Chen, Y. Xiang, M. Sun, Anticipating accidents in dashcam videos, in *Asian Conference on Computer Vision* (2016), pp. 136–153
5. R. Wagner-Kaiser. Using neural networks to detect car crashes in dashcam footage. https://blog.insightdatascience.com/crash-catcher-detecting-car-crashes-in-video-9c0522b04558. Last accessed 04 Oct 2019
6. CrashCather by rwk506. https://github.com/rwk506/CrashCatcher. Last accessed 13 June 2019
7. N. Shakla, *Machine Learning with Python* (Manning, New York, 2018)
8. D. Rosebrock, Deep learning for computer vision with python, in *PyImageSearch* (2017)
9. J. Howse, P. Joshi, M. Beyeler, *Computer Vision Projects with Python* (Packt Publishing, Birmingham, 2016)
10. S. Nikolenko, A. Kadurn, E. Arhangelskaya, Deep learning. Diving into world of neural networks. [In Russian: Glubokoe obuchenie. Pogruzhenie v mir neironnyh setei] (Piter Publishing House, Saint Petersburg, 2015)
11. Serious autocrash has happen nearby Lipetsk. [In Russian: Seryoznoe DTP sluchilos pod Lipetskom]. http://www.lipetskmedia.ru/news/view/115566-Syeryeznoye_DTP.html. Last accessed 22 May 2019

Smart City Platform Based on Citizen Reporting Services

George Suciu, Lucian-Alexandru Necula, Vladimir Jelea, Dragos-Sebastian Cristea, Carmen-Catalina Rusu, Luigi-Renato Mistodie, and Marius-Petrut Ivanov

1 Introduction

The smart city concept integrates different types of Internet of Things (IoT) devices for collecting data and using it to manage resources efficiently and improve the quality of life for the citizens [1]. Usually, collected data are analysed in order to improve traffic and transportation systems, utilities infrastructure, waste disposal, crime detection, community services and other services within the city [2]. The main factors that generated interest in the Smart City domain are environmental protection, ageing population, increase of urban population and economic restructuring [3]. The inhabitants of a city are increasingly looking for a friendly city, a smart city where people enjoy the services offered and the connections with the other inhabitants. By designing intelligent systems and connecting the existing ones, urban services address various needs and enhance the visibility of the inhabitants' problems to the authorities [4].

To address the main challenges of the Smart City domain, the CitiSim platform [5] is defined by its services, protocols and instruments for the development of smart services, 3D visualization and control of the Smart City ecosystem, and

G. Suciu (✉) · L.-A. Necula · V. Jelea
Beia Consult International, Bucharest, Romania
e-mail: george@beia.ro; lucian.necula@beia.ro; vladimir.jelea@beia.ro

D.-S. Cristea · C.-C. Rusu · L.-R. Mistodie
"Dunarea de Jos" University of Galati, Galati, Romania

Altfactor, Galati, Romania
e-mail: carmen.rusu@ugal.ro; luigi.mistodie@ugal.ro

M.-P. Ivanov
Altfactor, Galati, Romania
e-mail: marius.ivanov@altfactor.ro

D. Cagáňová et al. (eds.), *Advances in Industrial Internet of Things, Engineering and Management*, EAI/Springer Innovations in Communication and Computing,
https://doi.org/10.1007/978-3-030-69705-1_5

simulation for substantiation and implementation of strategic decisions within urban environments. CitiSim focuses on various domains of the city including mobility, health, comfort, security, emergencies, energy and citizen reporting.

In the context of the CitiSim project, the current paper presents a Citizen Reporting smart service meant to enhance the relationship between citizens and authorities and to solve problems with a higher efficiency within the Smart City. The rest of the article is structured as follows: Sect. 2 presents Smart City solutions and focuses on citizen reporting services, Sect. 3 depicts the CitiSim concept, Sect. 4 proposes a Citizen Incident Reporting smart service developed within the CitiSim project, while Sect. 5 concludes the paper.

2 Related Work

The current section presents various Smart City solutions that provide various functionalities for enhancing the living conditions of citizens. Solutions that aim to address various aspects of the Smart City ecosystem are oriented to provide custom tools for the city managers on request or to provide generic services that can be implemented in any city.

2.1 Smart City

The current sub-section presents solutions created to provide Smart City services to the citizens and society, thus improving aspects related to comfort, environment, health, traffic, utilities and economy.

The CitySDK [6] project provides a development environment for the creation of Smart City services using APIs (Application Program Interfaces). The solution has been tested in eight European cities, namely Amsterdam, Barcelona, Helsinki, Istanbul, Lamia, Lisbon, Manchester and Rome.

IBM Intelligent Operation Center [7] represents a platform provided by IBM which offers various Smart City services customizable on-demand. The challenges of the Smart City are addressed through data visualization and analytics which are the basis for optimizing operational efficiency. KPIs (Key Performance Indicators) facilitate the enhancing of city services, personnel and Smart City applications (traffic control, emergency management, incident reporting, etc.).

Sentilo [8] is a Smart City open-source platform with a modular architecture that facilitates its expansion without modifying the core system. The data visualization and management rely on a REST interface. Through the catalogue and administration console, the user can manage connected devices and associated users. At its origin, the solution was oriented towards the development of Barcelona in the Smart City domain, but the concept extended to cities as Terrassa and Reus.

Huawei Smart City [9] is a platform for real-time data reporting and analysis based on technologies as Cloud Computing, Internet of Things (IoT), Big Data and Artificial Intelligence (AI). It can integrate applications and devices in order to monitor the land planning, economy, transportation, population and ecological environment and manage specific operations related to emergencies, surveillance, utilities, traffic and parking, public illumination, urban agriculture, etc.

SmartCity Budapest Transport [10] is a mobile application for public transport information in Budapest. The application is free and can be used offline. This solution offers a public transport route planner, the full schedule of all the public transport lines of Budapest and a vector-based map with address search.

Smart City Traveler [11] is a solution that improves the experience of a person who likes to travel the city and wanted to explore the city by specifying the time in hours. Based on a questionnaire, the system analyses and creates a user program based on the time specified by the user for a trip.

SMART CITY Inspector [12] analyses RGB data, DSM, DTM or thermal data from drones and aircraft captured images. The application offers various solutions for smart data analysis, water management and smart planning for the future of cities. This solution is recommended for spatial planners, city builders, environmentalists, etc.

2.2 *Citizen Incident Reporting*

Citizen incident reporting applications allow citizens to effectively participate in the local governance by documenting their concerns and sending a report to the government order management services. The current sub-section presents solutions that aim to facilitate the communication between citizens and Smart City stakeholders through reporting of various events that may affect the ecosystem.

CitySourced [13] is a citizen reporting solution that allows the user to take a picture using the mobile phone camera, classify it in a category of interest and submit the report using a mobile application to municipalities or other stakeholders. Due to the usage of GPS to associate the report to a certain location, the need to add the address in the mobile application interface is not necessary.

Citizen Problem Reporter [14] represents a reporting application that allows the user to submit problems not addressing emergencies in their community using a mobile device or a personal computer. The report is used by the local government or other stakeholders to enhance the quality of citizen services. Thus, after analysing the requests, the local government can plan specific actions and send the request further to responsible parties for resolution.

MyTown Smart City [15] is an application used to report a problem, for example, a pothole, a street light outage, parking concerns or sensitive information anonymously via the HotLine tool. This application allows citizens to easily connect with local governments.

FixMyStreet [16] is an application made to be used online and offline. This application is used by citizens to report street issues. In this way, problems can be solved more quickly by the authorities.

At Romanian level, various Smart City reporting solutions are available for the citizens to report events. e-Alba Iulia [17] is a proximity mobile application that guides the user in exploring landmarks and to participate in events and provides the option to report issues that impact the city to local authorities. My Cluj [18] is an application provided by the Cluj-Napoca City Hall that provides a mobile service for reporting events associated with the public domain. Based on the user's location, images and the event information, the report is sent to the responsible public institutions for evaluation and intervention. A similar solution, Cluj Now [19] focuses on reports related to infrastructure work that affects urban traffic. The application also provides information on events affecting urban traffic for the public. Târnăveni Smart City [20] is another solution for reporting infrastructure problems (leakage of gas or water, dangerous driving conditions, etc.) to the municipality, thus lowering the intervention time of intervention for public authorities. The mobile application also provides information regarding such events and news related to infrastructure enhancements. With the Smart City Giurgiu application [21], the user can report various problems or irregularities that are encountered in Giurgiu, such as: asphalt pits, damaged benches in parks, public lighting issues, sewerage, etc. The reported problems are sent automatically to the Giurgiu City Hall, which will notify the competent authority. The user is notified by email about the status of the reported problem.

3 CitiSim Concept

In the last two decades, the proliferation of ICT technologies has led the evolution towards a new digital era that facilitates the sharing of information between citizens and environments at a global scale. With a continuous increase of urban population, cities are facing various challenges that can be managed using technology innovations [22]. In this context, a considerable number of research projects and pilots have been deployed as an attempt to study and manage resource consumption, environmental quality and urban mobility. However, due to a wide variety of technologies which pose various limitations in terms of scalability and integration, the proposed solutions imply a considerable effort for their administration and are not viable from an economic perspective.

Currently, the absence of a common platform that supports the implementation and integration of smart city services represents a blocking factor in the deployment of sustainable solutions. To address this challenge, the CitiSim ecosystem provides powerful monitoring and management solutions that enable city planners to substantiate viable decisions both on tactical and strategic levels based on the data collected from various sources. For a natural interaction with the Smart City ecosystem within the digital world, CitiSim provides 3D visualization instruments, facilitating the exploration of various augmented and virtual reality scenarios.

CitiSim provides a framework for a facile development and deployment of Smart City services, addressing various market verticals, such as energy management, environmental monitoring, urban mobility and emergency management. The main CitiSim instruments are:

- A platform consisting of different software blocks and data transmission protocols for Smart City services implementation;
- A 3D visualization tool for digital interactions with the Smart City ecosystem;
- A simulation framework for decision-making, successfully demonstrated in emergency management and energy efficiency scenarios as part of the CitiSim project.

Since smart services applications are moving from laboratories to the real city, there exists a growing need for advanced tools to manage and promote the smart city ecosystem. For the development of the so-called smart city services, aspects as advanced platforms (e.g. middleware, standards, protocols, interfaces, etc.), data visualization, and a safe software environment where testing new applications become capital issues. Consider an SME that is in charge of developing an application for a smart city. This SME currently finds several barriers: firstly, the access to the city information, such as the layout of pedestrian routes or people mobility patterns, that has to be elaborated by them. Secondly, the SME developer team has to deal with a variety of protocols, heterogeneous data format and data sources; and thirdly, there is a lack of environments for testing smart city applications. These barriers impact on time-to-market of feasible solutions and make very complicated to develop applications for smart cities and, in turn, the realization of smart cities.

The problem with the current state-of-the-art technologies is the lack of a holistic platform that helps developers to implement applications and to provide citizens with a powerful tool to translate data into meaningful information. Until today most of the products of the market, and research and innovation projects were focused mainly on developing specific applications or standardizing interfaces in the case of middleware. Simulation of different processes involved in the day-to-day of a smart city represents a challenge to be addressed in order to support tactical and strategic decisions about mobility, city layout design, emergency situations, etc. The CitiSim project aims to fill this gap and proposes an e-infrastructure that enables fast design and development, a safe test environment and the deployment of advanced applications for smart cities. This e-infrastructure is a key element in the smart city development path, which will provide to different agents (e.g. city councils, SMEs, and citizens) with a catalogue of advanced services built from up-to-date data that represents the state of the city.

Therefore, the final goal of CitiSim is providing a platform integrating multiple smart services that offer structured information collected from the urban environment, and whose knowledge is valuable for a wide range of stakeholders (see Fig. 1): firstly, citizens may take advantage of up-to-date, rich information on dimensions not explored before; SMEs may interact with CitiSim platform for developing pioneer, added-value services that help them to position them as leaders in relation to their competitors; large companies act both providing technology (e.g. hardware devices, network providers) and as developers of services and applications (e.g. data

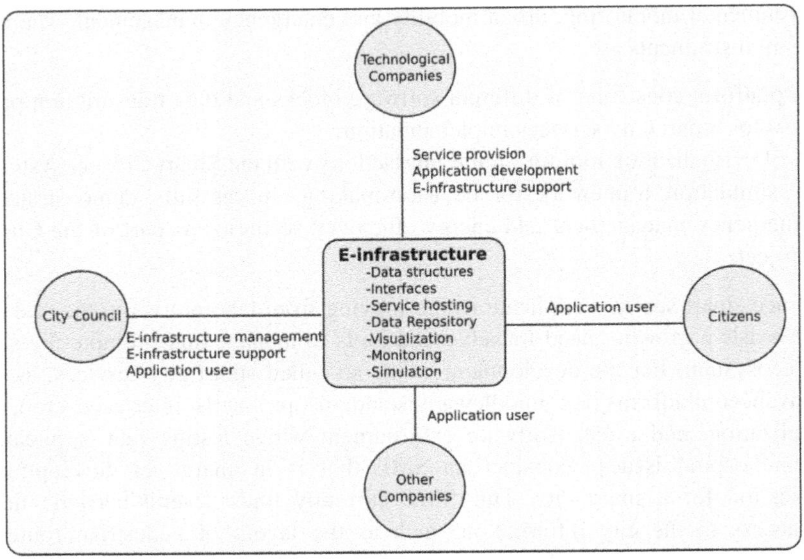

Fig. 1 Stakeholders of the market value chain for CitiSim

analysis, 3D models, algorithms) for the final users and, city councils hold most part of the CitiSim infrastructure, participate in its definition through regulations, contribute with their information systems, and act also as managers, obtaining strategical information from the city and from citizens.

This novel approach allows us focusing on the problem and to identify market-specific and common services with great potential of reusability. Two are the markets to which these uses cases are oriented, such as environmental friendly mobility and energy efficiency.

Figure 2 identifies seven steps involved in the production of smart services and their delivery to the end-users of the four markets cited above as well as their actors and enablers:

- Monitoring devices: The CitiSim infrastructure will be populated by a large set of heterogeneous devices for the monitoring of variables of interest for each particular use case. We are getting in touch with different hardware manufacturers (sensors, actuators, RFIDs) that may provide the most adequate devices.
- Smart things: Due to the interest showed by several city councils around the Europe (e.g. Rivas VaciaMadrid), the consortium will offer possibilities of collaboration to those organisms. To this purpose, they will provide requirements. This fact demonstrates that the CitiSim project already has an access market.
- Network integrator: Development and integration of networked devices that use different communication technologies.

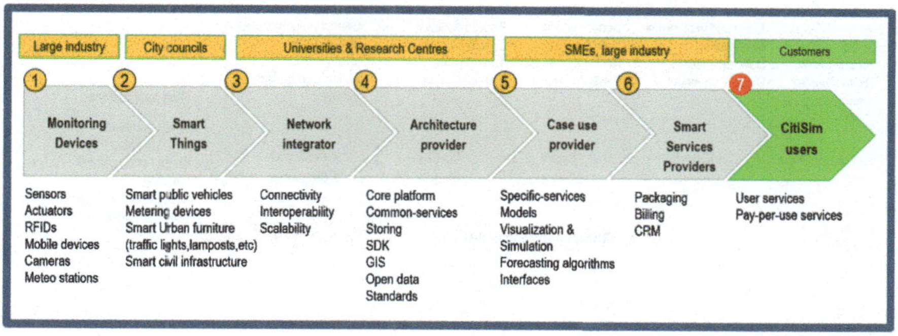

Fig. 2 Market value chain for CitiSim project

- Technology provider: Universities and research centres in collaboration with industrial partners in the CitiSim consortium will define the reference architecture for CitiSim based on the study of the state-of-the-art and their expertise.
- Use case providers: CitiSim partners grouped in a country-basis will propose at the beginning of CitiSim project the definition of a use case to be deployed on CitiSim architecture, and with relevance and potential impact on the citizens.
- Smart services providers: Smart models go beyond the urban models based on LUTI, cellular automates and agents found in the current state-of-the-art. The analysis of data by means of Big Data analytics about the city will allow defining smart services with added value for citizens. Advanced visualization techniques (augmented virtuality and augmented reality) need 3D models about different physical infrastructures (transport, energy, water, etc.), geodata visualization, and gamification.
- Project innovations and technology value chain.

The technological value chain is composed by a set of components where different technologies are involved and where partners of CitiSim accumulate proved experience:

- The monitoring and simulation smart city platform covers functional requirements described above and operational issues like security, scalability, Big Data software solutions, etc.
- Smart city low-level infrastructure: data proceed from different sources as heterogeneous multi-platform devices (sensors and actuators, GPSs, RFIDs, city infrastructure, smartphones) and social networks must be extracted, stored in a robust, scalable middleware and clouds, combined together, analysed and integrated.
- Smart urban models about different physical infrastructures and interest dimensions of the city (transport, energy, etc.) that are continuously and dynamically acquiring data from the things of the city, recomputing the outputs and forecasting behaviours in a mid-term predictive context. Smart models go beyond the urban models based on LUTI, cellular automates and agents found in the current

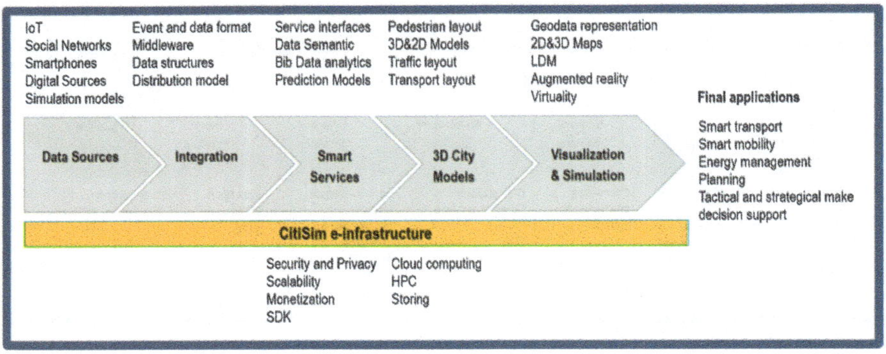

Fig. 3 Technology value chain for CitiSim project

state-of-the-art. The analysis of data by means of Big Data analytics about the city will allow defining smart services with added value for citizens.

- 3D visualization and simulation: advanced visualization techniques (augmented virtuality and augmented reality) need 3D models about different physical infrastructures (transport, energy, etc.), geodata visualization and gamification.

The technological value chain of CitiSim project is shown in Fig. 3.

The core of CitiSim is the development of a smart city ecosystem that provides both functional and non-functional issues and makes smart services available to citizens. The CitiSim reference architecture is presented in Fig. 4, and it is organized on five horizontal layers from bottom to up.

The five horizontal layers are presented within the following paragraphs.

- Cognitive Monitoring Layer: This layer is responsible for collecting data through the smart things of the CitiSim ecosystem. It is also more ambitious than the monitoring solutions found in the current state-of-the-art since it includes pre-processing modules for extracting cognitive data in addition to the monitoring itself of the smart city. The format of the data should be compliant with a standardized and open-data format, e.g. SensorML. This layer provides events to the rest of the layers of the e-infrastructure related to what and where is happening on the city.
- Simulation Layer: Simulation models are designed to use both the data of the Persistence Layer (synthetic data and data provided directly by the monitoring devices). According to the use-cases defined, CitiSim is going to put focus (without preventing the use of other models) on the following simulation models: people mobility model, traffic mobility model, mass transit model and business location model. This layer can provide to the service layer with raw events on the same way that of Cognitive Monitoring Layer. According to the desired configuration, a service/application can receive information from the simulation layer exactly on the same way that from the monitoring layer.

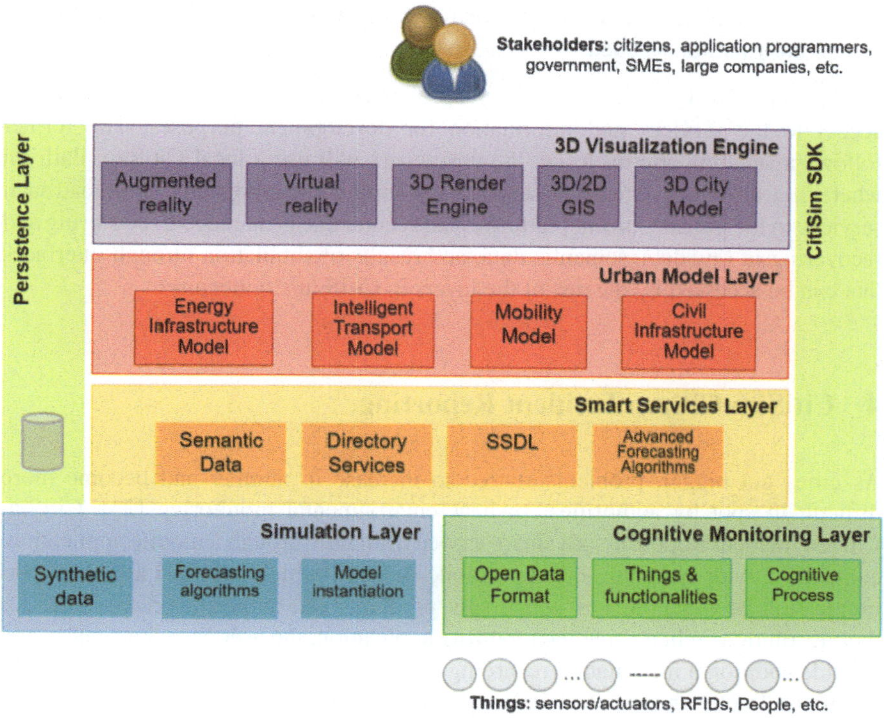

Fig. 4 Detailed reference architecture of CitiSim platform

- Smart Services Layer: This layer is the core of the e-infrastructure. Common services for several use cases can be reused by overlaying applications, such as directory services, services to provide semantic to data, and services description (Smart Services Definition Language, SSDL). From a developer point of view, these services provide with a well-known interface useful functionality like subscription mechanism to specific type of data related with, for example, people mobility, traffic mobility, etc. Specific middleware-related services (e.g. event distribution service or service composition service) are also included on this layer.
- Urban Model Layer: This layer provides all type of city layout urban models as a service. The models can be contributed by companies and citizens on a collaborative effort done by the smart city community.
- 3D Visualization Engine Layer: This layer uses the underlying layers for representing the city and geo-positioning the information of the city. Attending to the purpose of the final application, the data can be represented on a 3D virtual world, a 2D map, or on an augmented reality app, over a variety of devices. In the case of mayor console application, i.e. a holistic decision support system for municipalities, the application enables to control and monitor all parameters of

the city at real time on augmented reality world, that is, a 3D virtual world where real information is added.

CitiSim SDK Layer and Persistence Layer are vertical layers that enable the access to the services and information for development purposes. The former follows a sandbox approach, i.e. the developers will use a local CitiSim platform where they develop and test new services and applications and then they upload such services to the global CitiSim platform. The second one is intended to be storing and recovering of raw data, semantic data, and results of simulation through interfaces that can be accessed by the rest of the layers in CitiSim architecture.

4 CitiSim Citizen Incident Reporting

As cities got bigger, problems started to increase in number and become more difficult to spot by authorities, such as environmental monitoring [23]. CitiSim Citizen Incident Reporting is a smart service available through a mobile application, compatible with Android and iOS mobile devices, which provides a better communication between citizens which spot problems and authorities responsible for their resolution. When using the reporting application, the user can take a photo and provide additional information regarding the event. The photo is analysed using AI (Artificial Intelligence) and metadata describing the object is generated (building, road, car, tree). After the report is submitted, the event data are processed, stored in the database and displayed on a heatmap. In addition, based on the data provided by the user, the report is sent to the authorities responsible for managing that specific issue. Based on the location, the user can be informed regarding various events happening in an area of interest (construction work on the road, water leakage on the sidewalk, car blocking public transportation, etc.).

The CitiSim Citizen Incident Reporting application uses an API to send and receive JSON objects containing account-resource, report-resource and user-resource data. The map is implemented using Google's Maps SDK over which a heatmap is generated to show places with different densities of reports using different gradients.

The application uses an API consisting of three main parts:

- Account Resource
- Report Resource
- User JWT Controller

Using the Account Resource package, the user can register, login, post and get personal data like first name, last name, username, user id, latitude and longitude. The Report Resource package is used for managing reports. It requests report data such as the report title, image, latitude, longitude, time and assign it to the user that sends it.

The user can view the submitted reports as a list containing all the necessary data. In the same way, authorities or other stakeholders can view the received reports as a list and filter their content based on the metadata and urgency. For example, a report containing metadata such as "dangerous leak" will be treated as an emergency and investigated immediately by the responsible authorities. When the reported problem is acknowledged, addressed or solved, the user is notified through the application interface or by email.

The application communicates with the API using JSON Objects, these are data structures written in key/value pairs surrounded by curly braces. The keys and values are separated by a column and each key/value pair is separated by a comma. These objects are created by transferring all the necessary data for the API in a HashMap and then send it to the Cloud using the Volley library. HashMap is a collection type in java, it is used in this context for its efficiency when working with large data sets because elements are being searched using their HashCode. A JSON Object Request contains a request method (GET/POST), an URL (where the data is being sent), the data and a header containing the authentication token. The AI agent analyses the photo submitted by the user, generates metadata associated with the identified objects or events and sends the data to the database. The CitiSim Citizen Incident Reporting service architecture is presented in Fig. 5.

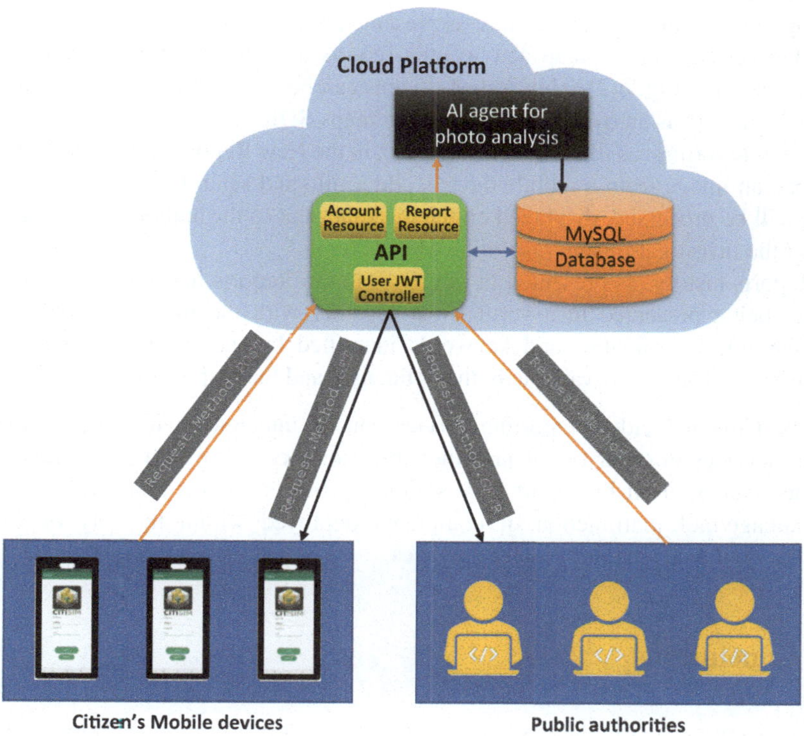

Fig. 5 CitiSim Citizen Incident Reporting service

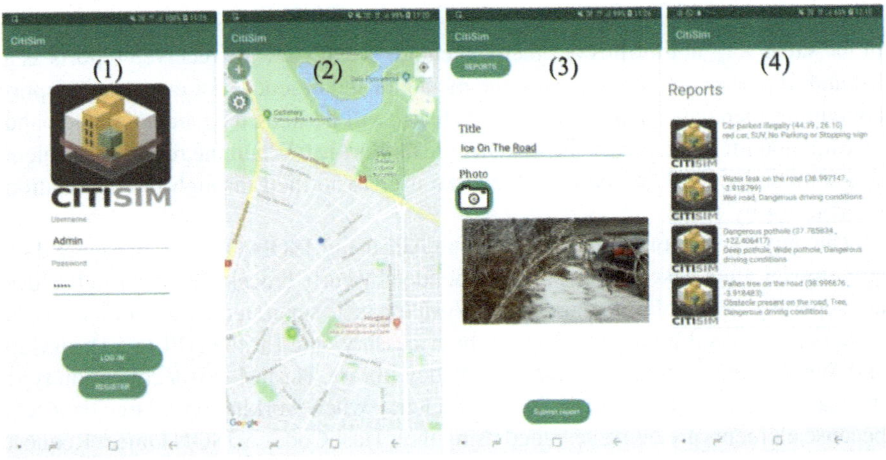

Fig. 6 Citizen incident reporting mobile application GUI

The Citizen Incident Reporting application provides the following functionalities:

- Log in activity: in the Log in screen the user is able to register to the platform or log in using an existing account, as presented in Fig. 6(1).
- Map activity: in the map activity, the reports are illustrated as a heatmap, as presented in Fig. 6(2). Thus, other citizens can be aware of the events in an area of interest. It is an overlay on the Google maps SDK.
- Event reporting: as illustrated in Fig. 6(3), in the New Report activity the user can load an image from a mobile device, add a title and send it to the server where it will be processed, displayed on the map and sent to the authorities responsible for the investigation and resolution of the event.
- Reports history: as presented in Fig. 6(4), in the Reports list activity the reports are being presented in a vertical list together with the title and latitude and longitude coordinates, and keywords identified by the AI agent after image analysis. The list is available both for the user and the responsible authorities.

The Citizen Incident reporting service can be integrated with other available smart services in order to enhance their functionality. For example, the reporting service can be used by Smart City solution providers to facilitate the reporting of damage/theft/malfunction of equipment deployed within the city or issues encountered when using the smart services.

5 Conclusions

The concept of Smart Cities is based on urban development by integrating systems and technologies to more easily manage the resources of a city and to improve the livelihood of its residents. The goal of a Smart City is to promote technical innovation and to improve the efficiency of urban management.

Mobile applications may be an important part in the process of improving a city by facilitating the communication between citizens and the authorities responsible for the city administration. To address the difficulties in reporting various events within the city, the CitiSim Citizen Incident Reporting application facilitates actuation towards problem-solving by using AI to categorize the event (e.g. damage to a monument, water leakage affecting traffic, parked car affecting the tram, ice on the road), send it to responsible authorities, and inform citizen about events in an area of interest and their resolution.

With this application, the authorities can monitor all the problems in the city and can find solutions to ensure a better quality of life for the citizens. As future work, we intend to enhance the performance of the AI agent in terms of object/event recognition. Another aspect would be providing the possibility to upload specific to damage or malfunction of the CitiSim devices deployed within the Smart City or the application available for the users.

Acknowledgements This work has been supported in part by UEFISCDI Romania and MCI through projects CitiSim, ESTABLISH, PARFAIT and WINS@HI, funded in part by European Union's Horizon 2020 research and innovation program under grant agreement No. 826452 (Arrowhead Tools), No. 787002 (SAFECARE), No. 777996 (SealedGRID) and No. 813278 (A-WEAR).

References

1. P. Ta-Shma, A. Akbar, G. Gerson-Golan, G. Hadash, F. Carrez, K. Moessner, An ingestion and analytics architecture for IoT applied to smart city use cases. IEEE Internet Things J. **5**(2), 765–774 (2017)
2. A.K. Singh, D. Kumar, V. Prakash, Importance and needs of IoT in developing smart cities (2019)
3. S. Myeong, Y. Jung, E. Lee, A study on determinant factors in smart city development: An analytic hierarchy process analysis. Sustainability **10**(8), 2606 (2018)
4. F. Bhatti, M.A. Shah, C. Maple, S.U. Islam, A novel internet of things-enabled accident detection and reporting system for smart city environments. Sensors **19**(9), 2071 (2019)
5. CitiSim Homepage, http://citisim.org. Last accessed 15 May 2019
6. CitySDK Homepage, http://citysdk.waag.org. Last accessed 14 May 2019
7. IBM Intelligent Operation Center Homepage, https://www.ibm.com/ro-en/marketplace/city-insights. Last accessed 14 May 2019
8. Sentilo Homepage, http://www.sentilo.io/wordpress/. Last accessed 19 Mar 2019
9. Huawei Smart City, https://e.huawei.com/en/solutions/industries/smart-city. Last accessed 19 Mar 2019
10. SmartCity Budapest Transport on Google Play, https://play.google.com/store/apps/details?id=hu.ponte.mobile.smartcity. Last accessed 19 Mar 2019
11. Smart City Traveler on Google Play, https://play.google.com/store/apps/details?id=com.smartcitytraveler. Last accessed 14 May 2019
12. Smart City Inspector on Google Play, https://play.google.com/store/apps/details?id=com.smart_city_inspector. Last accessed 14 May 2019

13. CitySourced Homepage, https://www.esri.com/news/ArcUser/0111/files/citysourced.pdf. Last accessed 15 Apr 2019
14. Citizen Problem Reporter Homepage, https://solutions.arcgis.com/local-government/help/citizen-problem-reporter/. Last accessed 15 Apr 2019
15. MyTown Smart City on Google Play, https://play.google.com/store/apps/details?id=com.mytown.realtermenergy. Last accessed 15 Apr 2019
16. FixMyStreet Homepage, https://play.google.com/store/apps/details?id=org.mysociety.FixMyStreet. Last accessed 15 Apr 2019
17. e-Alba Iulia on Google Play, https://play.google.com/store/apps/details?id=ro.gebs.discover.albaiulia. Last accessed 6 Mar 2019
18. My Cluj on Google Play, https://play.google.com/store/apps/details?id=com.indsoft.mycluj&hl=en. Last accessed 6 Mar 2019
19. Cluj Now, https://play.google.com/store/apps/details?id=ro.lifeishard.clujnow. Last accessed 6 Mar 2019
20. Târnăveni Smart City on Google Play, https://play.google.com/store/apps/details?id=unserver.tarnaveni. Last accessed 6 Mar 2019
21. Smart City Giurgiu on Google Play, https://play.google.com/store/apps/details?id=ro.itom.smartcitygiurgiu. Last accessed 6 Mar 2019
22. Á. Carrera, E. Merino, P. Aznar, G. Fernández, C.A. Iglesias, An agent-based simulation model for emergency egress, in *International Symposium on Distributed Computing and Artificial Intelligence* (2018), pp. 140–148
23. G. Suciu, M.A. Sachian, A. Pasat, LoRa architecture for air quality monitoring, in *Advanced Topics in Optoelectronics, Microelectronics, and Nanotechnologies IX*, vol. 10977, (International Society for Optics and Photonics, Bellingham, WA, 2019), p. 109770T

Creativity of Different Managerial Generation Groups in the Slovak Industrial Enterprises

Natália Vraňaková, Dagmar Cagáňová, and Andrea Chlpeková

1 Introduction

Creativity has always been at the heart of business [1]. Organizations and enterprises are pushing in order to solve the problems that have arisen through creative methods. Therefore, creativity is becoming as an important tool. Government businesses also need to adapt to new technologies, as threats require ingenuity and creativity [2].

Creativity is an important part of managerial work and overlaps with all management functions. Creative thinking is therefore part of planning, organizing, leading people and making decisions. A number of scientific articles and books are written about the issue of creativity. Various authors have discussed the mentioned issue and therefore it is necessary to define the characteristics of creativity.

Studies conducted in the previous period have focused on creativity as a way of problem as an option of solving problems requiring the interaction and integration of general knowledge of specific areas, skills in scientific processes and different scientific creativity [3]. The author Konigova [4] defined creativity as the most manifestation of life. Creative thinking is a prerequisite for original ideas, where we engage the imagination, apart from bounding reality and strive for new insights both on ourselves and on social issues or professional and scientific issues. According to the author Muzik [5], creativity is the ability of a person to be active, creatively receive new knowledge and be productive in the mental area. Without creativity, solving a problem that requires more than the reaction itself and learned action would be completely unthinkable.

N. Vraňaková · D. Cagáňová (✉) · A. Chlpeková
Slovak University of Technology in Bratislava, Faculty of Materials Science and Technology in Trnava, Institute of Industrial Engineering and Management, Trnava, Slovakia
e-mail: natalia.vranakova@stuba.sk; dagmar.caganova@stuba.sk; andrea.chlpekova@stuba.sk

D. Cagáňová et al. (eds.), *Advances in Industrial Internet of Things, Engineering and Management*, EAI/Springer Innovations in Communication and Computing,
https://doi.org/10.1007/978-3-030-69705-1_6

The authors of the scientific monograph chapter agree with the author Konigova who defines creativity with an emphasis on creating original ideas and resolve any questions. Creativity can be seen from a variety of perspectives, as it is based on the characteristics of a person. Currently, there are four generation groups at the labour market. Each generation group has grown up under different conditions and therefore has different characteristics, positives, negatives and expectations from the employer.

Representatives of the Baby boomers generation group are characterized by their years of birth between 1946 and 1962. In general, they are hardworking and their position, benefits and prestige act as a motivating factor. They also like long working weeks and often point to their work success. Members of mentioned generation group are confident and independent. As the generation group grew up in a period of frequent reforms, representatives are convinced that they could change the world. The oldest generation group is deeply focused on success and career and they are characterized by commitment, finesse, ingenuity and the desire to win in every situation [6].

Next in line is generation group X, which is characterized by the fact that it is currently the second oldest generation group on the labour market and they were born between 1963 and 1977. According to the author Kane [7], the mentality of the generation group X members is greatly influenced by the shift of the manufacturing economy towards the service economy. Finesse in the technological skills is due to the fact that they were first to grow up with the computers and thus technology is an integral part of their lives. Compared to the other generations, members of generation group X are more willing to change jobs to achieve their goals. They are also willing to adapt to change and are relaxed towards an alternative lifestyle. Representatives of generation group X are in general very ambitious, willing to learn new things and acquire new skills. On the other hand, they like to do things according to their own ideas. Managers who represent generation group X often use humour and games in their work activities.

Generation group X is followed by the third generation group on the labour market, Millennials or generation group Y. The Millennials were born between 1978 and 1992. According to the author Kane [8], generation group Y is the fastest-growing segment in the labour market. At present, enterprises tend to compete for talented people, so managers cannot ignore and neglect the needs, desires and attitudes of the most widespread generation. Each generation group is characterized by certain traits, which are also caused by the conditions and culture in which they grew up. Technologies are the most common characteristic of generation group Y. They rely on technology and also seek support in it in order to carry out their work more efficiently. Members of generation group Y use smartphones, laptops and other technological means through which they are available 24/7. The most popular means of communication include e-mail, text messages and communication via new social media platforms such as Twitter or Instagram. In this way, they communicate not only with friends, but also with colleagues, because without Internet or smartphone they cannot imagine the world. Generation group Y likes to work in a team. One of the disadvantages of mentioned generation group is the fact that they are often looking for new and better jobs. It is no exception that they stay in the enterprise for only a few years and during that time they do not have enough time to work their way up to another better position.

The last generation group Z is currently also the youngest generation group on the labour market. The members were born between 1992 and 2010. Francis and Hoefel [9] identified members of generation Z as digital natives. From an early age, they were greatly influenced by the Internet, social networks and mobile system. Due to the mentioned technological conveniences, from generation Z grew up a comfortable generation in the collection of information and their subsequent verification. They feel a strong need for the free speech and also want to understand all types of people. Members of generation group Z are financially literate, they feel the importance of saving money for the future. With this fact, they consider the stability and security of the job as important and the salary is at the second place. They also prefer full-time work compared to part-time work, which was a surprising finding compared to generation group Y.

Based on the study of theoretical background [10–13], the authors of the scientific monograph chapter have created Table 1, which gives an overview of generational groups and their common features and differences.

As various generation groups work side-by-side at the workplace, enterprise's behaviour should change in order to adapt to each generation [14]. According to the author Murphy [10], there are several benefits of the multigeneration work team:

1. The team has the ability to attract and keep talented people of all age categories.
2. The team tends to be more flexible.
3. The members of multigeneration team tend to gain and keep the larger market share through the members, which reflect a multigeneration market.
4. Decisions are stronger and right because they are based on several aspects.
5. The team has the ability to be innovative and also more creative.
6. The team with people from various age categories is reflecting the needs of a diverse public and can achieve more effective cooperation.

As it is evident from the characteristics of each generation group, the authors consider as important to look at the issue of managerial creativity not only in terms of hierarchical status, but also in terms of belonging to the generation group.

2 The Current Situation of Employees on the Labour Market in the Slovak Republic

Each generation group is characterized by its qualities, positives and negatives. For this reason, the authors of the scientific monograph chapter have looked at the issue of creativity from the perspective of generational groups. The authors also consider it as crucial to analyse the current labour market situation in terms of age categories and changes in their structure.

Table 1 Common features and differences of generation groups

	Baby boomers	Generation X	Generation Y	Generation Z
Work and work ethics	Workoholic, quality work	Work as hard as needed, flexibility, meaningful and independent work	Meaningful work, enterprising approach	Meaningful and independent work
Preferred learning style	One-on-one coaching, classroom instruction, peer interaction and feedback	One-on-one coaching, assessment and feedback, discussion groups	Discussion groups, assessment and feedback, need to see context and value	Online education, interactivity in the classroom
Work-life balance	Did not know about work-life balance. Sacrificed personal life for work	Spend more time with the family. Appreciate work-life balance	Work-life balance is needed—they came with this idea	Work-life balance is necessity—they will do nothing to the detriment of the family
Preferred method of communica-tion	Phone calls, personal interaction	Voice mail, Email	Text messages, Email	Instant messages, social networks
Motivation	Have affection to be motivated by leaders who get them engaged and show them how they can attain a difference	Allow them to get the job done on their own schedule	Have affection to be motivated when their managers connect their acts with their personal and career aims	They are motivated by job freedom, flexible job forms and laissez-faire style of management
Rewards	Personal appreciation, promotion, and recognition	Free time, upgraded resources, opportunities for development, bottom-line results, certifications	Awards, certificates, tangible evidence of credibility	Financial rewards and career advancements
Developmental areas	Computer training, leadership	Skills training in areas of expertise, team building	Problem-solving, decision making	Interpersonal relationship, communication skills

Over the next 10 years, the proportion of older employees in Europe will increase. The above-mentioned fact is due to the demographic changes (longer life expectancy and low birth rates). The European workforce will be older and up to 30% of the employed will be the older employees [15].

In Table 2, it can be seen the distribution of labour force (in thousands) by age category and years in the Slovak Republic [16–19].

For better clarity, the authors of the scientific monograph chapter have decided to process the table into a graphical form (Fig. 1).

Table 2 and Fig. 1 show that the number of employees in older age categories is increasing. This increase is steeper compared to employees in younger age categories. Compared to 2015 and 2018, the number of employed in the oldest age category increase by 10 thousand. The age category of employees 55–64 has the steepest increase in employment between 2016 and 2017, particular 29.1 thousand. In the age category from 45 to 54, the number of employees increased by 22.9 thousand compared to 2017 and 2018. Similarly, employees in the age group from 35 to 44 recorded an increase of 22.5 thousand between 2015 and 2016. The age category that recorded a decrease of employees is a category from 25 to 24. In 2016, 626.6 thousand were employed in the mentioned age category. In 2018, however, it was only 609.6 thousand and in this case, there was a decrease of 18 thousand employees. Also, the age category from 15 to 24 recorded a decrease, especially between 2016 and 2017 by 14.2 thousand.

The authors of the scientific monograph chapter have transformed the age categories of employees into generational groups. Figure 2 represents generational employee groups in 2018.

Based on Fig. 2, there can be seen that currently there are four generational groups on the labour market in the Slovak Republic. Almost half of all employees are employees of generation X. The second-largest group are the employees of generation Y. The youngest generation in the labour market, generation Z accounts

Table 2 Employed by age category and year (own elaboration, 2019)

	15–24	25–34	35–44	45–54	55–64	65+
2015	155.2	608.8	701.3	570.3	337.7	18.6
2016	151.7	627.6	723.8	589.9	349.5	19.1
2017	165.9	613.8	736.9	591.3	378.6	26.2
2018	155.8	609.6	742.7	614.2	391.9	28.6

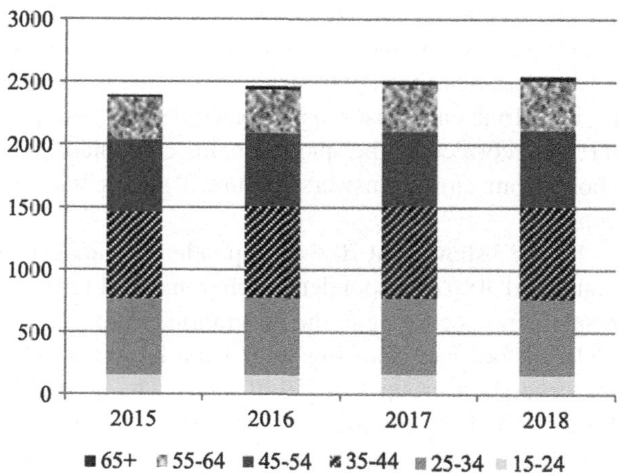

Fig. 1 Overview of employees by age category and years. (Own elaboration, 2019)

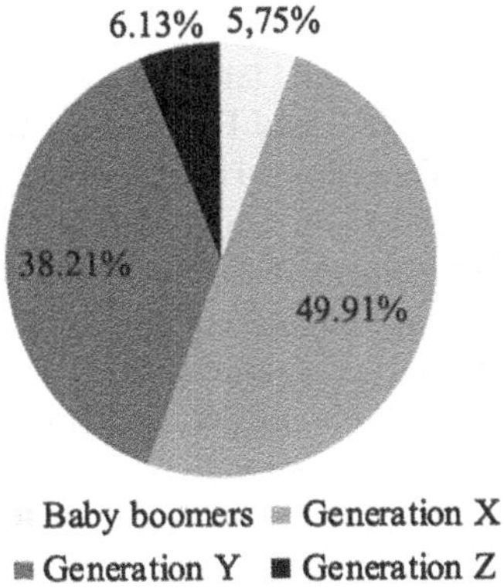

Fig. 2 Generation groups of employees in the Slovak Republic. (Own elaboration, 2019)

for 6.13%. The least employees are members of the oldest generation group, Baby Boomers, who make up 5.75% of all employees in Slovakia.

3 Materials and Methods

The aim of the research was to investigate and characterize the perception of creativity and its importance among managers of different generation groups in industrial enterprises in the Slovak Republic.

3.1 The Research Sample and Data Collection Tools

The research sample consisted of managers of industrial enterprises from the Slovak Republic. The research was carried out in April 2019. The industrial enterprises were in the field of automotive (Tier 2) and electrotechnical production. The questionnaire was physically distributed to enterprises. In total, 42 managers of industrial enterprises participated in the research. Forty questionnaires were returned correctly. The questionnaire contained eight questions and respondents chose from closed answers. Figure 3 shows the composition of respondents by gender.

Figure 3 shows that 70% of respondents (managers in industrial enterprises) were males and 30% of respondents were females. Figure 4 shows the composition of the respondents according to the generation group.

It can be seen from Fig. 4 that the largest generation group of respondents is the generation group Y (35% of total). The second largest group is generation X, followed by Baby boomers. At least, 10% are respondents from generation group Z. It is clear that the composition between generation groups is unequal, but the authors can say that the representation of respondents corresponds to the composition

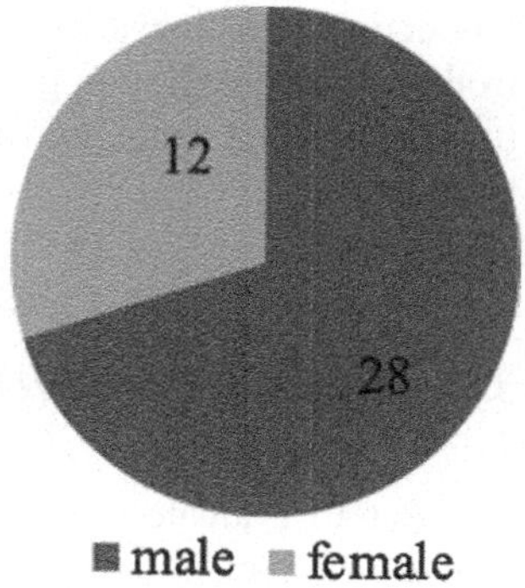

Fig. 3 Composition of respondents according to the gender. (Own elaboration, 2019)

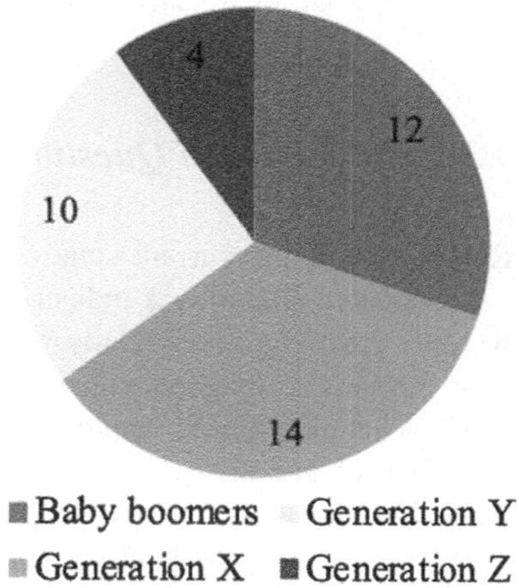

Fig. 4 Composition of respondents according to the generation group. (Own elaboration, 2019)

Table 3 Composition of respondents according by hierarchical status and generation group (own elaboration, 2019)

Generation group/type of manager	First line manager	Middle line manager	Top manager	Cumulative
Baby boomers	8	3	1	12
Generation X	5	8	1	26
Generation Y	8	2	0	36
Generation Z	3	1	0	40
Sum	24	14	2	

of generation groups on the labour market. Table 3 shows the composition of respondent by hierarchical status and generation group.

It is clear from Table 3 that respondents from generation group Z are, for the most part, first-line managers (e.g. team leader, head of department or head of the working team). It can be assumed that especially due to their young age and lack of work experience, they are not part of top management (board member, plant manager, president, vice president). A similar situation exists with the generation group Y. The largest part of middle line managers (e.g. economic, personnel, technology, production, logistics) are respondents from the generation group X. One respondent from mention generation group belongs to the top management. A similar situation (one respondent from top management) exists with the generation group Baby boomers.

3.2 The Research Methods

In order to evaluate the collected data, the authors of the scientific monograph chapter used quantitative statistical methods, such as frequencies histograms, pie charts and additional analysis in table form.

3.3 The Research Question (RQ)

RQ1: What is the current state of perception of the importance of creativity by managers in the Slovak industrial enterprises according to the generation group?
RQ2: Is creativity part of daily work of managers from different generation group in the Slovak industrial enterprises?
RQ3: What methods do employers support the creativity of managers in the Slovak industrial enterprises?
RQ4: Do managers of different generation groups in the Slovak industrial enterprises have an interest in developing their creativity?

4 Results and Discussion

The evaluation of research questions:

RQ1: What is the current state of the important perception of creativity by managers in the Slovak industrial enterprises according to the generation group?

In order to evaluate the first research question, the authors of the scientific monograph chapter have used the fourth question from the questionnaire: "In my opinion, creativity is an important part of every manager".

The respondents had the opportunity to choose one answer from five offered: 1—strongly disagree, 2—disagree, 3—neutral, 4—agree, 5—strongly agree. In the authors' opinion, it is important to analyse how the individual generation groups perceive creativity, especially due to their differences.

Figure 5 shows respondents' answer according to generation groups.

It follows that 25% of the respondents from generation Baby boomers considers creativity as unimportant. Creativity is the most perceived by the respondents from generation X. The respondents from generation Y and generation Z agree with the opinion that creativity is very important. The authors of the scientific monograph

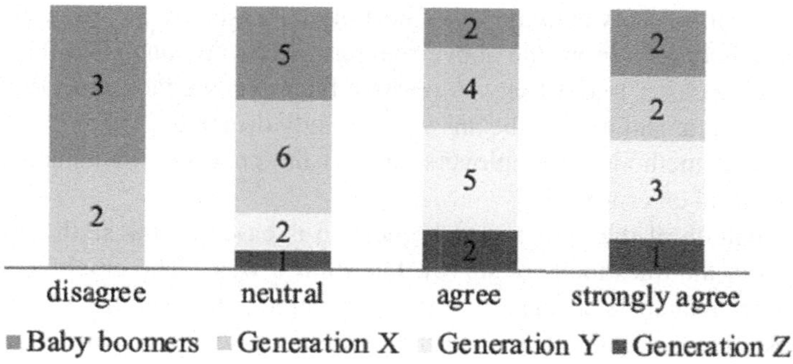

Fig. 5 Perceiving the importance of creativity according to the generation group. (Own elaboration, 2019)

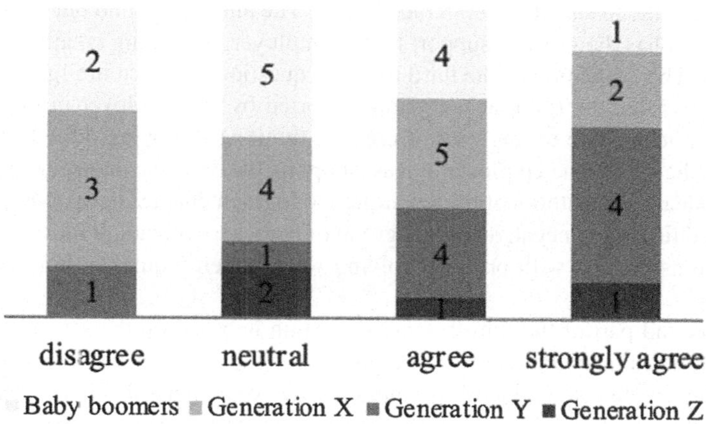

Fig. 6 Creativity as part of my daily work according to the generation group. (Own elaboration, 2019)

chapter consider as positive that neither respondent mark the answer "strongly disagree".

RQ2: Is creativity part of daily work of managers from different generation group in the Slovak industrial enterprises?

In order to evaluate the second research question, the fifth question from the questionnaire was used: "Creativity is part of my daily work". The respondents were asked again to choose an answer from the offered range: 1—strongly disagree, 2—disagree, 3—neutral, 4—agree, 5—strongly agree. The authors of the scientific monograph chapter also have decided to evaluate the second research question from the perspective of the generation group of managers in order to find out which generation uses creativity most often in their work. The evaluation of the second research question can be seen in Fig. 6.

It follows that the youngest generation group, generation Y and generation Z use the creativity in daily activities in the highest rate. The "disagree" answer in terms of creativity as part of daily activities was most often marked by the generation group X and Baby boomers. From the above findings, it can be concluded that younger generations are closer to creativity. A positive finding is the fact, similarly to RQ1, that neither respondent market the answer “strongly disagree”.

RQ3: What methods do employers support the creativity of managers in the Slovak industrial enterprises?

The evaluation of the third research question is based on the sixth and seventh question from the questionnaire survey. Question 6 was: “The employer supports me and other managers in making creative thoughts/ideas/solutions”. As in previous questions, the respondents had the opportunity to choose one answer from five offered: 1—strongly disagree, 2—disagree, 3—neutral, 4—agree, 5—strongly agree.

The third research question is not evaluated by the generation group, but according to the manager’s hierarchical status. The aim was to find out which group of managers has the greatest support by the employer in creating creative ideas and solutions. The evaluation of the third research question is shown in Fig. 7.

It follows that the top managers are supported by the employer in the creation of creative ideas, due to the need of creative thinking following directly from the nature of the work. The employer at least supports the first-line managers in making creative ideas. The authors of the scientific monograph chapter have considered the mentioned finding as negative, as this group of managers is leading most people and it is often associated with problem-solving which often requires a high degree of creativity.

The second part of the third research question is based on the seventh question from the questionnaire survey: “If your employer supports you and other managers in making creative thoughts/ideas/solutions, by what method?” The respondents had the opportunity to choose from eight answers (brainstorming, SWOT analysis, synectics, mind map, bank of ideas, Delphi method, flow method, other). Table 4 shows the evaluation of responses.

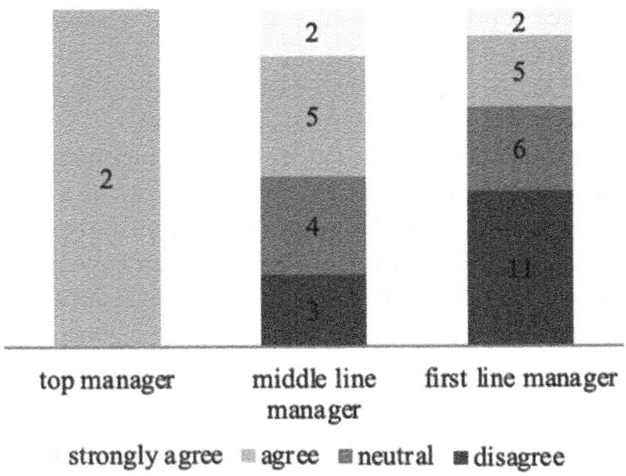

Fig. 7 Promoting managers’ creativity by employers. (Own elaboration, 2019)

It is clear from Table 4 that the most utilized method for supporting creativity is the brainstorming method. Brainstorming is followed by mind map and SWOT analysis. As the area for improvement, the authors consider the development of the method "bank of ideas" to the practice of the industrial enterprises, as it represents an extension of the brainstorming method and it is advisable to use the mentioned method in unusual situations where anonymity is appropriate.

RQ4: Do managers of different generation groups in the Slovak industrial enterprises have an interest in developing their creativity?

The fourth research question is based on the last, eighth question from the questionnaire: "If the employer gave me the opportunity to take part in a course/training/workshop aimed at developing my creativity, I would participate in this activity". The respondents once again chose the answer on scale 1—strongly disagree, 2—disagree, 3—neutral, 4—agree, 5—strongly agree. The respondents' answers are shown in Fig. 8.

Figure 8 represents that the youngest generation groups—generation Y and generation Z have the greatest interest in creativity training and creativity development workshops. Part of the generation X and generation Baby boomers marked that they would not be interested in training and workshops. In the authors'

Table 4 Methods for supporting creativity (own elaboration, 2019)

Method/type of manager	Top manager	Middle line manager	First line manager
Brainstorming	1	3	2
Swot analysis	0	2	1
Mind map	0	1	3
Bank of ideas	0	0	1
Other	1	1	0

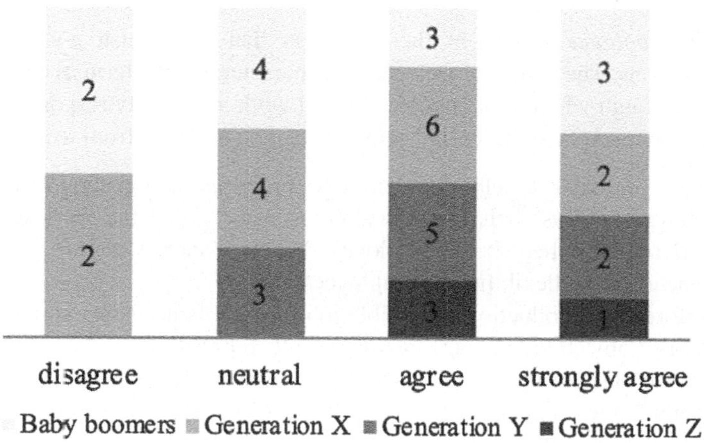

Fig. 8 Interest in developing creativity according to the generation group. (Own elaboration, 2019)

opinion, this finding should be an incentive for the employer. Employers should emphasize the importance of creativity for older employees in the enterprise. It is the older generation groups that create employees who transfer their knowledge and experience to younger generations, and creativity is an essential part of these activities.

In the context of the managers' different generation groups creativity, the authors of the scientific monograph chapter also have considered the managing multigenerational work teams as very important. On one hand, the only creative manager can manage multigeneration teams. On the other hand, a manager who manages subordinate employees in an appropriate style can also create a working environment which encourages employees to think creatively and to make creative and innovative ideas.

According to the author Hannam [12], there are five several areas which include the following measures to focus on successful multigeneration management:

1. *Communication*
   (a) There is a need to build awareness of the four generations presence benefits. Communicate about diversity. The goal is the effective communication in different ways for creativity and innovation. The unique skills of each generation can unleash hidden potentials.
   (b) Speak and communicate in a way that motivates each generation. Effective communication is needed for all generation groups at the workplace. Unclear matters can have long-term consequences, especially for inexperienced employees.
   (c) Author Murphy [10] also includes in the field of communication the need of initiating conversations about generations. When the negative aspects of individual generations get out, they become less personal and general. It is easier to talk about them.
2. *Work-life balance* is one of the indicators that distinguish good and great organizations. The author of Delecta [20] characterized the term as an individual ability, through which it is possible to fulfil work and family responsibilities, as well as other responsibilities and activities that do not arise from working life.
   (a) Work-life balance helps to retain top employees who generate creative and innovative ideas. It is necessary to reconsider what the work is, when it is done and where it can be done. Managers should review how may be concerned the flexibility for employees.
   (b) Before the introduction of flexible job forms it is necessary to have clearly defined objectives and performance results for each job.
   (c) On the basis of the author Tolbize [11] finding, the members of generation group X most likely want to work-life balance, member of age groups also aspire to the same thing.
3. *Growth and development*—each generation groups has its own method for processing information and the author Tolbize [11] stated that members of

all generation group value personal growth and personal gratification. For generation group X and generation group Y it is less likely that they would prefer reading guides. Rather, the mentioned method is favoured by the generation of Baby Boomers.

(a) The younger generation needs to be involved in meaningful projects to satisfy their desire for meaningful and important work. Members of generation group Y want to make a change and they will leave, if they feel undervalued.
(b) Some companies have banned social networks at the workplace. However, wherever it is possible, it is appropriate to use social networks to attract, engage and retain the best talents.
(c) Through mentoring and employee development, employees can enhance and improve their performance and learn new skills. Mentoring programs that connect leaders and managers with the best employees are suitable.

4. *Providing recognition and rewards*—according to the author Hansen [21], the terms recognition and reward are often considered as synonyms. However, they are different and also they are essential tools for achieving the employee satisfaction.

(a) The recognition and reward system needs to be reviewed in order to be not outdated and meets the need of all generation group of employees. In the interest of quality system, it is necessary for employees to define what they really expect.
(b) It is useful to offer a system of rewards and recognition and to diversify it according to generation groups. The system should reflect different priorities and motivating factors. The list of rewards can include the simplest rewards and employees should choose rewards based on their preferences.
(c) Many enterprises also appreciate employees for the small things that employees do. The awards should reflect both the culture of the enterprise and the generational preferences of the employees.
(d) Recognition of employees should be an important part of each manager and should be included as part of performance evaluation.

5. *Employee engagement*—author Ologbo [22] characterise employee engagement as an important factor for every business. Engaged employees contribute to business success and commitment is reflected in customer service. As a result, committed employees help create loyal customers.

(a) Engaged employees are needed for the enterprise and they can inspire other employees. Engaged, productive and talented employees need to be noted. Large proportion of managers focuses their attention on employees who are average or below average.
(b) It is appropriate to transform average or part involved employees into committed employees.

Collaboration of different generational groups in the workplace can bring its advantages and disadvantages. These are the older employees who should transfer their experience and knowledge to the younger ones. On the other hand, younger employees should help older colleagues in technology and modern trends. Such collaboration also provides a quality work environment in which employees feel comfortable and are also creative and innovative. In Table 5 are shown the strengths and weaknesses of older employees according to the author Svobodová [23]:

Older employees have also opportunities and threats [23], shown in Table 6.

A creative and innovative manager should be able to take advantage of the benefits of older employees and eliminate the negative aspects. It is also necessary to focus on the opportunities that come with older employees. However, enterprises must also be aware of the threats and try to prevent them as much as possible.

Table 5 Strengths and weakness of older employees [23]

Strengths	Weaknesses
Experience, practice, expertise	*Mental productivity*
Experience, practice, expertise, skill, knowledge, knowledge in the framework of already forgotten skills (fur processing, textile blending, etc.)	Psychophysical changes (in making decision, concentration), deteriorating sensory perception (sight, hearing), loss of potential, short-term memory, bad habits, stereotypes
Characteristics	*Motivation*
Reliability, opinion stability, responsibility, thoroughness, care, prudence, perseverance, patience	Loss of self-esteem, phlegmatism, lower willingness to professionally educate, low motivation for performance, low work commitment
Relation to the work performed	*Performance*
Knowledge of problems, knowledge and orientation in the field, effort to give the best work performance, self-realization	Reduced rate of physical performance and mobility, increased fatigue, stress, decreased productivity
Relationship to the company	*Health condition*
Loyalty, business loyalty, job stability, low turnover	High morbidity, health restrictions, more health problems, transition to disability
Contacts	*Adaptability*
Knowledge of a given group of companies, companies and individuals for making contacts of work importance	Less creativity, distrust of news, routine, worse adaptability, worse adaptation to new methods
Priorities	*Relationship to the new technique*
Clear priorities, possibility to devote to work fully (grown up children), time flexibility, enough time	Lower computer literacy, unwillingness to learn, lower levels of internet literacy
Relations, dealings	*Relations, dealings*
Mature approach to people, managing emotions, good interpersonal communication, prudence	Reduced tolerance, explosiveness, inadaptability
	Other
	Unwillingness to teamwork, less ability to keep up with younger employees

Table 6 Opportunities and threats of older employees [23]

Opportunities	Threats
Job forms	*Job opportunities*
Working part-time, working weekends, calm work pace	Fewer job opportunities, age discrimination, fear of job loss, fast pace
Use of experience	*Costs*
Exploitation of life and professional experience, exploitation of strong knowledge potential	Higher societal costs, morbidity, occupational accidents, disability
Sharing experiences	*Employee/employer relations*
Transfer of knowledge and experience, reciprocal exchange with younger employees	Non-existent systems of working with older employees, not interested in exploiting the potential of older employees
Retraining, education	*Relations between older and younger employees*
Courses, revitalization programs, lifelong learning, support for professional growth	Intergenerational tension, competition in the form of young colleagues, young management
New technologies	*Health care*
The possibility of using technology if older workers have the will to learn new things	Failure to provide optimal medical care in case of illnesses, failure to provide adequate care
Other	*Presentation of elders*
Longer vacation state support for the retention of older workers, retirement opportunities for active retirees	Stereotyped thinking about employees aged 50+

5 Conclusion

The main aim of the scientific monograph chapter was to investigate and characterize the perception of creativity and its importance among managers of different generation groups in industrial enterprises in the Slovak Republic.

The analysis and the realization of the questionnaire survey confirmed that there are currently four generation groups on the labour market—generation Baby boomers, generation X, generation Y and generation Z.

The questionnaire survey conducted some negative findings. The authors of the scientific monograph chapter can conclude that 12.5% of respondents do not consider creativity as an important part of every manager. 15% of respondents do not consider creativity as part of their daily work. On the contrary, positive findings include the fact that up to 40% of managers are supported by the employer in creating creative ideas, and up to 62.5% of the respondents are interested in training and workshops in the field of creativity development.

The authors of the scientific monograph chapter have considered as very important that employers support employees in creativity and give them opportunities to further develop their creativity, as well as author Lenhardtova et al. [24] and Hornak et al. [25]. Manager's creativity level affects two upper levels the same way as the enterprise's creativity level affects the creativity level of employees and of a specific manager [26]. Mentioned issue is still relevant, as evidenced by the number of articles and books on creativity. Due to the current demographic situation, the authors of the scientific monograph chapter consider the need to look at creativity also from the perspective of generation groups. All four-generation groups are currently meeting at the workplace and the employer should adapt the conditions to ensure their satisfaction. The analysis has shown differences between generations, which proves the need to approach to generation groups individually and also according to author Caganova et al. [27] constantly monitor their opinions and needs.

The authors of the scientific monograph chapter have considered all the findings from the questionnaire to be important in relation to the industrial enterprises practice.

Acknowledgement The scientific monograph chapter is a part of VEGA project No. 2/0077/19 "Working competences in the context of Industry 4.0".

References

1. T.A. Amabile, K. Mukti, *Creativity and the Role of the Leader* (Harvard Business School Publishing, Cambridge, MA, 2008)
2. S. Williams, Increasing employees' creativity by training their managers. Ind. Commer. Train. **33**(2), 63–68 (2001)
3. K.K. Yang et al., Exploring the significant predictors of convergent and divergent scientific creativities. Think. Skills Creat. **31**, 252–261 (2019)
4. M. Konigova, *Creativity: Techniques and Exercises (Tvorivost: Techniky a Cviceni)* (Grada Publishing, Praha, 2007)
5. J. Muzik, *Andragogical Didactics (Andragogicka didaktika)* (Codex Bohemia, Praha, 1998)
6. S. Kane, Baby boomers in the workplace, https://www.thebalancecareers.com/baby-boomers-2164681. Last accessed 2019/4/5
7. S. Kane, The common characteristics of generation X professionals, https://www.thebalancecareers.com/common-characteristics-of-generation-x-professionals-2164682. Last accessed 2019/4/5
8. S. Kane, The common characteristics of millennial professionals, https://www.thebalancecareers.com/common-characteristics-of-generation-y-professionals-2164683. Last accessed 2019/4/5
9. T. Francis, F. Hoefel, True Gen: Generation Z and its implications for companies, https://www.mckinsey.com/industries/consumer-packaged-goods/our-insights/true-gen-generation-z-and-its-implications-for-companies. Last accessed 2019/4/5
10. S.A. Murphy, *Leading Multigenerational Workforce* (AARP, Washington, 2007)
11. A. Tolbize, *Generational Differences in the Workplace* (University of Minnesota, Research and Training Center on Community Living, Minneapolis, MN, 2008)
12. S. Hannam, B. Yordi, *Engaging a Multi-Generational Workforce: Practical Advice for Government Managers* (IBM Center for the Business of Government, Washington, 2011)
13. A. Turner, Generation Z: Technology and social interest. J. Indiv. Pyschol. **71**(2), 103–113 (2015)
14. B.A. Kaifi, W.A. Nafei, N.M. Khanfar, M.M. Kaifi, A multi-generational workforce: Managing and understanding millennials. Int. J. Business Manag. **7**(24), 88–93 (2012)

15. J. Ilmarinen, *Promoting Active Aging in the Workplace (Podpora aktivneho starnutia na pracovisku)* (Europska agentura pre bezpecnost a ochranu zdravia pri praci, Bratislava, 2012), https://osha.europa.eu/sk/publications/promoting-active-ageing-workplace. Last accessed 2019/4/10
16. Statistical Office of the Slovak Republic, *Q1 Results of the Labour Force Sample Survey 2015* (Vysledky vyberoveho zistovania pracovnych sil za prvy stvrtrok, Statisticky urad Slovenskej republiky, 2015), ftp://193.87.31.84/0206754/Vysledky_vyberoveho_zistovania_pracovnych_sil_za_1_stvrtrok_2015.pdf. Last accessed 2019/4/10
17. Statistical Office of the Slovak Republic, *Q1 Results of the Labour Force Sample Survey 2016* (Vysledky vyberoveho zistovania pracovnych sil za prvy stvrtrok, Statisticky urad Slovenskej republiky, 2016), https://www7.statistics.sk/PortalTraffic/fileServlet?Dokument=0c1140b1-4d10-4ae1-bcfa-e8be6502f379. Last accessed 2019/4/10
18. Statistical Office of the Slovak Republic, *Q1 Results of the Labour Force Sample Survey 2017* (Vysledky vyberoveho zistovania pracovnych sil za prvy stvrtrok, Statisticky urad Slovenskej republiky, 2017), https://slovak.statistics.sk/PortalTraffic/fileServlet?Dokument=71ad7534-cea3-496a-8b5f-be445be01090. Last accessed 2019/4/10
19. Statistical Office of the Slovak Republic, *Q1 Results of the Labour Force Sample Survey 2018* (Vysledky vyberoveho zistovania pracovnych sil za prvy stvrtrok, Statisticky urad Slovenskej republiky, 2018), https://slovak.statistics.sk/PortalTraffic/fileServlet?Dokument=3f33a5ee-c977-488b-b12e-77bd839cdc07. Last accessed 2019/4/10
20. P. Delecta, Work life balance. Int. J. Curr. Res. **3**(4), 186–189 (2011)
21. F. Hansen, M. Smith, R.B. Hansen, Rewards and recognition in employee motivation. Comp. Benefits Rev. **34**(5), 64–72 (2002)
22. C.A. Ologbo, S. Saudah, Individual factors and work outcomes of employee engagement. Proc. Soc. Behav. Sci. **40**, 498–508 (2012)
23. L. Svobodová, *Occupational Health and Safety and Risk Factors of Employees over 50 Years—Employment of Older People in the Work Process with Proposals for Measures at Sector and Company Level (BOZP a rizikove faktory pracovniku nad 50 let—uplatneni starsich osob v pracovnim procesu s navrhy na opatreni na urovni odvetvi a podniku)* (VUBP, Praha, 2009)
24. Lenhardtova, Z., Caganova, D., Gyurak Babelova, Z.: Contemporary situation in the creativity area in the Slovak republic. In: Annals of DAAM and Proceedings of DAAM Symposium 2009: 20, 1: Proceedings of the 20th International DAAM Symposium "Intelligent Manufacturing & Automation: Focus on Theory, Practice and Education", pp. 937–938, DAAM INT Vienna, Vienna (2009)
25. F. Hornak, D. Caganova, M. Cambal, The role of managerial creativity in human potential development, in *EAEEIE 2012 23rd EAEEIE Annual Conference*, (University of Cagliari, Cagliari, 2012), 4 p
26. Y. Raschshepko, Creativity of company and creativity of manager: How to measure them, in *Proceeding of the 4th International Conference of Management, Leadership and Governance (ICMLG 2016)*, (Acad Conferences Ltd., Curtis Farm, 2016), pp. 516–519
27. D. Caganova, D. Ruskova, Creativity in ESP, in *CO-MAT-TECH 2005 Proceeding/International Scientific Conference 13th*, (STU v Bratislave, Bratislava, 2005), pp. 166–169

Stages of the Virtual Technical Functions Concept Networks Development

Petro Bakhovskyy, Mykola Yevsiuk, Oleg Zabolotnyi, Dagmar Cagáňová, and Anatolii Tkachuk

1 Introduction

Considering mentioned above, it turns out that when implementing the concept of VTF, it is expedient to organize a centralized subscriber database for a particular network, which takes part in the process of using the corresponding virtual technical function. In this case, each switching node of mobile stations must have subscriber data MS, which now are in the zone of its service. When moving a mobile station from the service area of one MSC to the service area of another MSC, they will receive on request the necessary information from a centralized subscriber database that will have access to other databases. In addition, in order to avoid unauthorized access by intruders to the network and possible theft of information or equipment, it is necessary to ensure the ability to verify the access rights of service personnel to the network (authentication) and identification of the relevant equipment. Saving authentication data and a list of serial numbers of the equipment involved also requires a single network of databases, process and store the relevant information. The problems of interaction of mobile communication standards of different generations are becoming more and more urgent because their number has been improving for 10 years. Today the standards of 6-, 7-, 8- are considered, although in practice the technologies of 4-, 5- generations have not yet been fully mastered. The problems of radio frequency resource allocation, the uncertainty of frequency band distribution, and the lack of network infrastructure and subscriber

P. Bakhovskyy (✉) · M. Yevsiuk · O. Zabolotnyi · A. Tkachuk
Technological Faculty, Lutsk National Technical University, Lutsk, Ukraine
e-mail: a.tkachuk@lntu.edu.ua

D. Cagáňová
Faculty of Materials Science and Technology in Trnava, Slovak University of Technology in Bratislava, Trnava, Slovakia

D. Cagáňová et al. (eds.), *Advances in Industrial Internet of Things, Engineering and Management*, EAI/Springer Innovations in Communication and Computing,
https://doi.org/10.1007/978-3-030-69705-1_7

equipment remain unresolved. The rapid development of mobile communication means forces operators and providers to search for the most optimal options for the implementation of next-generation mobile telecommunications networks, which could ensure their further technical and economic functioning with the specified quality criteria. To do this, you must have a sufficient level of competence that would allow you to determine the features of such promising technologies. Taking this into account, the concept of the VHE concept was introduced [1, 2] providing for many options, both its own technical implementation and improvement of the concept itself. However, a wide range of options causes serious problems, requires a certain systematization, adaptation of existing networks operating in the telecommunications market and determining the prospects for their further development. The journal “Zviazok” (“Connection”) (no. 4/2010) shows some problems and difficulties in the deployment of future generation networks [3]. To continue the topic, the article highlights the common features and differences in the features of the VHE concept, reflected in various publications. The introduction of LTE technology allows operators to reduce capital and operating costs, reduce the total cost of ownership of the network, expand their capabilities in the field of convergence of services and technologies, and increase revenues from the provision of data services. The network supports MBSFN (Multicast Broadcast Single Frequency Network), which allows the introduction of services such as mobile TV as opposed to DVB-H.

Features provided by LTE:

- High network bandwidth
- High sensitivity
- Support for gaming applications due to low response time
- High interactivity
- Data download speed
- Voice over IP/IMS
- High quality of service
- Expansion of mobile TV channels
- High image quality of mobile TV
- A wide range of options for coexistence of LTE network and networks of previous generations

Comparing the wide range of advanced technologies developed in the technologically advanced countries with the conditions and resources existing in Ukraine, it is easy to conclude that there is a need for a constant careful selection of a limited number of new network technologies, which should focus the attention of all existing telecommunication network operators involved in the market of information and communication services in Ukraine (importers or manufacturers of equipment and software, system integrators, developers, entrepreneurs, regulators).

Given the openness of foreign policy and the market nature of domestic policy, there are practically no restrictions on the use of any network technologies in Ukraine. The only limiting factor is the ratio between the cost of funds for the introduction of technology and the purchasing power of the Ukrainian market of info communication services.

The above fundamental paradigms regarding the technical characteristics of the implemented projects for building mobile telecommunications networks lead to the conclusion that any scenario for the implementation of virtual technical functions with specified quality requirements can be implemented in the concept of SAE/EPS project networks.

Over the past decade, mobile telecommunications systems have become widely used. With the growth of effective demand for mobile telecommunications network services, it became necessary to expand their capabilities. Despite the variety of technologies and high quality of services provided by mobile communication systems of 3-, 4-, 5- generations, they are in principle not able to meet all the requirements of a certain category of users who need to have an ultra-wide range and ultra-high speed of data reception/transmission over global distances with the specified requirements. This category of users includes specialized structures that perform, in addition to providing normal services, also remote control services for certain process that require receiving/transmitting data today for today. These requirements include high capacity, transmission quality, and throughput. The inability to implement these requirements based on existing systems leads to further search for dynamic development of mobile devices. To meet this demand, it became possible only by creating a project of a switching network consisting of several nodes, within which any mobile station (MS) must have access to communication services. MS will be served by the mobile switching centre (MSC) in the service area of which it is located at a certain time. The system approach means that the mobile station is not included in a specific switching node, but directly in a co-existing network system. This feature significantly complicates the organization of the subscriber database, which is the main component in some scenarios of implementation of the concept of VTF (Virtual Technical Functions).

The complexity of the project of organization, monitoring and management of the subscriber database is explained by the fact that MS can move freely, changing the service areas of both base stations (BS) and MSC.

Taking into account the mentioned above and implementing the VTF concept, it is advisable to organize a centralized subscriber database for a particular network, which takes part in the process of forming the corresponding virtual technical function. In this case, each switching node of mobile stations must have subscriber data MS, which at this time are in the zone of its service. When moving a mobile station from the service area of one MSC to the service area of another MSC, these databases get the necessary information from a centralized subscriber database. In addition, in order to avoid unauthorized access of subscribers to the network and possible theft of equipment, it is necessary to ensure the ability to verify access rights (authentication) and identification of equipment. Storing authentication data and a list of hardware serial numbers also requires a network-wide database that stores the relevant information.

2 Literature Review

In networks of the international standard for mobile digital cellular communication of the GSM (Global System for Mobile Communications) standard and in its modifications routing in the mode of data transmission is carried out in the mode of channels switching containing certain shortcomings. The fact is that the Internet traffic is "pulsating" in nature, because the data are transmitted in sharp bursts, not a continuous stream. This type of data traffic is not well suited for circuit-switched networks due to the connection exists even when data are not being transmitted. In addition, it is very expensive for the end user, who pays for the time, not the amount of data transmitted and makes inefficient use of the operator radio interface.

The inconveniences also include a long connection establishment time and the fact that the connections are implemented at a time when the quality of transmission through the radio interface decreases significantly. Through these disadvantages, the transmission of switched packets over circuit-switched mobile telecommunications networks is unreliable and expensive.

In some works variants of construction of concepts are estimated by volume of the transmitted signal information. To theoretical study of the concepts paid considerable attention such scientists as M. Torabi and R. Buhrke [4].

The above sufficiently restrained the introduction of any concept as a video home system (VHS) and VTF.

Taking into account the above mentioned, a new standard was developed to eliminate these problems and to introduce the provision of Internet services in the mobile telecommunications network.

One of the important steps in the evolution of GSM networks was the introduction of packet transmission technology GPRS (General Packet Radio Service).

Based on the improved IMT 2000 project, new requirements for a unified system of mobile telecommunications were formulated [1, 5–9]. Thus, within the framework of the European telecommunication standards Institute ETSI (European telecommunications standards institute), five basic concepts of radio access projects for next-generation mobile telecommunications systems were considered. Two of the five technologies received the most valuable recognition:

- WB-CDMA (wideband code division multiple access)—for paired frequency bands;
- TD-CDMA (time division—code division multiple access)—for odd bands.

The European concept of mobile telecommunications the third generation project systems creation received the name UMTS (universal mobile telecommunications system) [1, 6–9].

The United States, in turn, submitted four projects, which were reduced to such options:

- the first option is based on further improvement of TDMA/AMPS technology (DAMPS, damage-associated molecular patterns);

- the second option offers to gradually increase the capacity of the CDMA One system, moving from the existing infrastructure to the CDMA 2000 technology;
- the third option—two proposals of the United States, which almost completely coincide with the proposals from Europe (UTRA, UMTS Terrestrial Radio Access) and Japan (WCDMA, Wideband Code Division Multiple Access) and in the process of further consideration merged into a single project.

In the Asia-Pacific region, two different draft standards for IMT-2000 were developed: CDMA and CDMA II. These were developed by the telecommunications technology association (TTA) of South Korea. In China, such a project is based on the use of a combined access method and a combination of TDMA and CDMA technologies.

In the framework of the project, IMT-2000 is valid for two migration strategies from existing networks in the mobile telecommunications networks of the project SAE/EPS:

1. Evolutionary
2. Revolutionary

The advantages and disadvantages of these two strategies are discussed in detail in a number of publications [1, 2, 6–10].

When adopting world standards for SAE/EPS project systems, two associations were formed, which were formed in the form of two partner associations:

1. 3GPP—Third Generation Partnership Project
2. 3GPP2—Third Generation Partnership Project 2

For the first time, these partnerships described the challenge of virtual technical functions concepts implementation. However, this description is general and does not contain the list of functions. To date, each of the telecommunications operators is the terminology of the description and implements the project independently.

The main contribution of the 3GPP partnership to the IMT 2000 project is the harmonization of five following projects:

1. UTRA-FDD
2. WCDMA
3. WCDMA NA
4. WIMS
5. CDMA II

The second partner project Association—3GPP2—is a supporter for the evolutionary development of mobile telecommunications 2G two directions in the technologies. The overall picture of the draft standards next-generation mobile telecommunications systems harmonization is shown in Fig. 1.

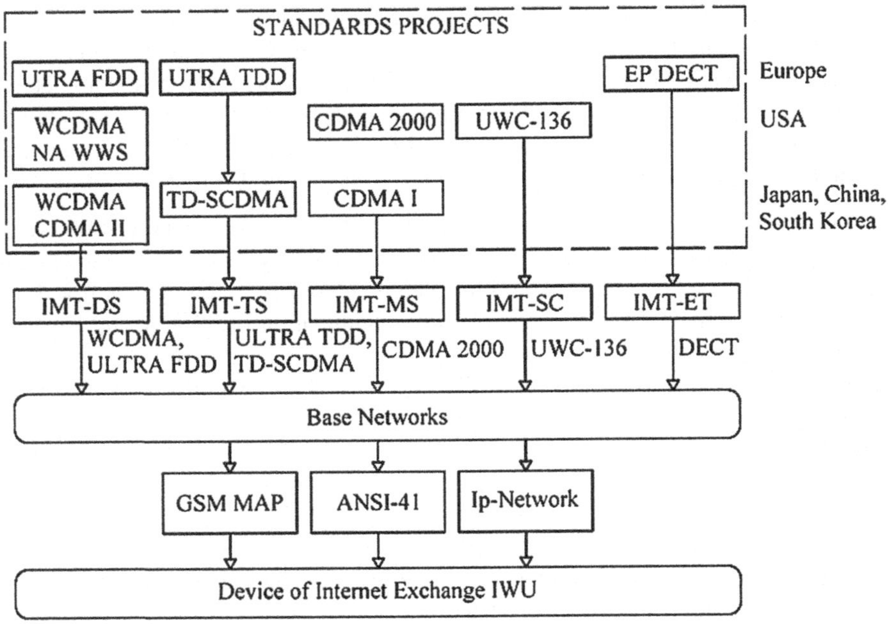

Fig. 1 The architecture of the terrestrial network IMT-2000. (Source: [2, 8–10])

3 Research Methodology

According to many experts, the implementation of SAE/EPS project systems will be preceded by a very long period of their joint existence with third-generation systems. With this in mind, to provide global roaming, as protocols for basic networks in the SAE/EPS project systems, it is planned to use advanced projects of 2G and 3G network protocols (GSM MAP and ANSI-41) and in the future—protocols based on IP-technologies and technologies of intelligent networks (IN) [1, 2, 6–11].

The purpose of this article is the analytical processing of mobile telecommunications networks implemented projects construction.

The above mentioned are the fundamental comments on the technical characteristics of mobile telecommunications networks implemented projects construction. These indicate the potential for the implementation of any virtual technical functions scenario with a given quality in the SAE/EPS project networks.

The authors of the paper analyses the technical capabilities of VTF technology implementation in SAE/EPS networks.

SAE/EPS services enable:

- View the web resources contents or send emails using phones and laptops that support HSPA (High-Speed Packet Access)
- Use HSPA modems instead of DSL modems
- Forward video and music using your existing 3G phones

The SAE/EPS project should be the next step in the development of mobile telecommunications systems. The GSM and WCDMA/HSPA standards are integrated into the SAE/EPS using standardized interfaces connecting the SGSN node (serving GPRS support node) and the enhanced reference network.

The existing concept of quality of service (QoS) for GSM and WCDMA systems is characterized by a sufficient level of complexity. However, the SAE/EPS project has a tendency to implement the concept of QoS, which would combine the simplicity and flexibility of access with support for backward compatibility. The SAE/EPS project proposes the use of a class-based quality of service concept. It provides a simple and effective solution for differentiating different package services. As a conclusion, the authors of the paper can state that the qualitative indicators of the implementation of the VTF concept will be provided by the implementation of the concept of quality of service based on classes.

According to the recommendations of ITU-T (International Telecommunication Union) developed within the IMT-2000 project, the SAE/EPS mobile telecommunications network consists of the following subsystems [2, 10, 12]:

- UIM subsystem (User Identity Module): provides user identification by the network, secure access procedures for both the subscriber and the network, and can perform various functions when performing additional services. UIM functions can be located on a separate card from the mobile terminal (MT) or can be physically integrated into the MT.
- MT subsystem: provides the ability to interact UIM and radio access network (RAN). Like UIM, the MT subsystem can perform various functions when performing additional services.
- Subsystem RAN: provides the possibility of interaction of the MT and the core network (CN) via radio section; can perform various functions when performing additional services.
- CN subsystem: provides the basic functions of the call service process and support for user mobility; can perform various functions in the performance of additional services.

The approach to the implementation of services in the SAE/EPS project, according to ITU-T recommendations, can be based on the concept of an intelligent network [2, 10, 13]. A fundamental requirement to the architecture of intelligent networks is the separation of the functions of service provision from the switching functions.

Each component of the SAE/EPS network participates in the implementation of a certain set of functional tasks and, if necessary, interacts with other functional units of its or other networks.

Each of the functional tasks can be represented as a set of smaller functions. Under the function, it should understand the set of the mobile telecommunications network action components in the process of services implementation once accessed. In this case, the various functions can be located in the same components. The obtained functions can be located in various components of both home (supporting) and visiting network. Thus, there is a set of variants (scenarios) of the VTF concept functional construction structures organization.

Functionally, the core of the SAE network includes four key components. The Mobility Management Entity (MME) module provides storage and management of service information about the subscriber, temporary identification data generation, terminal devices authorization in land-based mobile networks and General mobility management. The UPE subscriber management module (User Plane Entity) is responsible for downstream terminating, data encryption, routing and forwarding of packets. ("Anchor") 3GPP plays the role of a gateway between 2G/3G and LTE networks. Finally, the functions of the SAE anchor are similar to those of the previous component, but serve to support the continuity of service when moving a subscriber between networks that meet and do not meet the 3GPP specifications. Consequently, this indicates the potential for the proposed VTF concept implementation. The generalized scheme of SAE/EPS architecture is shown in Fig. 2.

In Fig. 2, the following notation is taken: Gb, lu, SGI, Rx+, $S1$–$S7$—interfaces; HSS—Home Subs Server; SGSN—Serving GPRS Support Node; PCRF—Policy and Changing Rules Function. 1—MME UPE; 2—EPS core; 3—"Anchor" SAE.

Modern development of mobile telecommunications systems and networks is marked by intensive development and implementation of next-generation systems-LTE-SAE (Long-Term Evolution-System Architecture Evolution) [4, 15, 16]. The

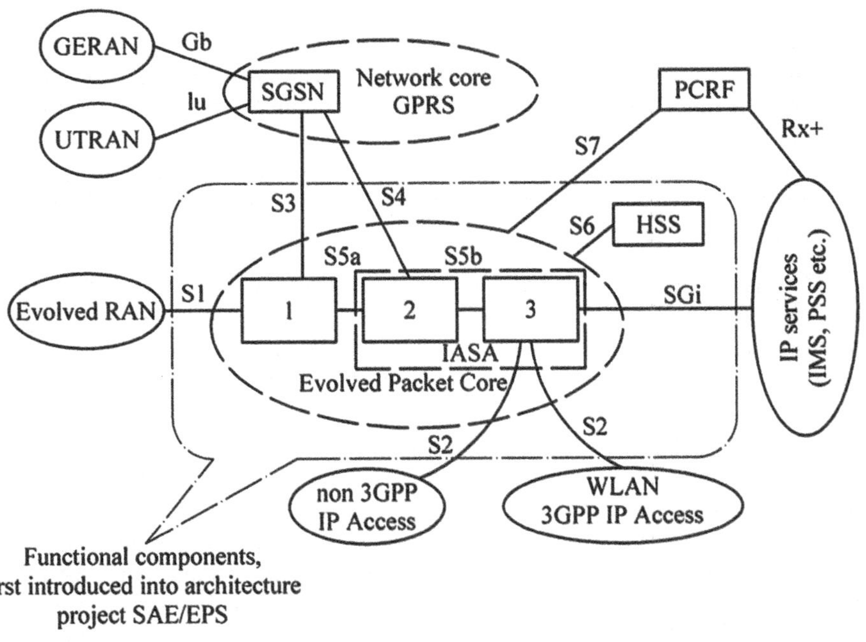

Fig. 2 The main components of the SAE/EPS project architecture. (Source: [14])

ever-growing migration of users between the networks of different Telecom operators requires the support of existing services in their business (home) networks. As an example, in 3G networks, this problem was solved within the concept of a virtual home environment-VHE (Virtual Home Environment), on the basis of which, in turn, the concept of virtual technical functions-VTF (Virtual Technical Functions) was built [7, 17]. The international organization 3GPP (Third-Generation Partnership Project), which initially developed improved versions of CDMA and UMTS, coordinates the research and development of LTE technology. 3GPP specifications defined ways to optimize interaction between networks such as 3G, SAE/EPS (EPS-Evolved Packet System), GSM and CDMA, which allows mobile operators to transfer their networks to new technologies [1, 18]. The beginning of work on the creation of SAE/EPS architecture specifications was due to the processes of transforming the infrastructure of mobile telecommunications networks into non-hierarchical packet switching networks based entirely on the IP Protocol. These works are carried out by the system architecture working group, which is part of the 3GPP consortium. The fundamental principles of the concept for SAE/EPS can be defined as follows:

- Improvement of basic performance indicators (connection time, voice quality, etc.), ensuring acceptable values of the total delay of data transmission over the network.
- Support for basic packet services, including VoIP and presence detection service.
- Support for existing and future access systems, taking into account the needs of the user and the technical policy of the operator, ensuring correct billing when the subscriber changes access technology.
- Support for access control functions to information resources (identification and authorization).
- Support for consistent QoS parameters throughout the network, especially when transmitting traffic between different domains and networks.
- Ensuring continuity of service for mobile subscribers;
- Ensuring confidentiality and protection of information when it is transmitted in the network.

According to ITU-T recommendations developed within the framework of the IMT-2000 project (International Mobile Telecommunications-2000), the SAE/EPS mobile telecommunications network consists of the following subsystems [8, 19]:

- UIM (User Identity Module) subsystem: provides user identification by the network, secure access procedures for both the subscriber and the network, and can also perform various functions when performing additional services. UIM functions can be located on a separate card from the mobile terminal (MT) or can be physically integrated into the MT.
- MT subsystem (Mobile Terminal): provides the possibility of interaction between the UIM and the RAN radio access network. Like UIM, it can perform various functions when performing additional services.

- RAN subsystem (Radio Access Network): provides the possibility of interaction between the MT and the base network (CN) via radio link; can perform various functions when performing additional services.
- CN subsystem (Core Network): provides the main functions of the call service process and support for user mobility; can perform various functions when performing additional services. The approach to the implementation of services in the SAE/EPS project, according to ITU-T recommendations, can be based on the concept of an intelligent network [9, 20]. A fundamental requirement for an intelligent network architecture is the separation of service delivery functions from switching functions. Each component of the SAE/EPS network participates in the implementation of a certain set of functional tasks and, if necessary, interacts with other functional units of its own or other networks. Each of the functional tasks can be represented as a set of smaller functions. A function should be understood as a set of actions of a mobile telecommunications network component in the process of implementing a service with a single access to it. In this case, various functions can be located in the same components. The resulting functions can be located in various components of both the home (support) and business network. Thus, there are many options (scenarios) for the organization of structures for the functional construction of the VTF concept. Functionally, the core of the SAE network includes four key components:
- The MME (Mobility Management Entity) mobility management module provides storage and management of service information about the subscriber, generation of temporary identification data, authorization of terminal devices in terrestrial mobile networks and General mobility management.
- The UPE (User Plane Entity) subscriber management module is responsible for downlink termination, data encryption, routing and packet forwarding. In this case, the "Anchor" 3GPP plays the role of a kind of gateway between 2G/3G and LTE networks. The SAE anchor functions are similar to those of the previous component, but serve to maintain service continuity when a subscriber moves between networks that meet and do not meet 3GPP specifications. The fundamental difference between mobile telecommunications networks of the 4G generation from mobile networks of previous generations is the ability to provide high-speed data transmission on the following radio stations:

  2048 Mbit/s—for office work
  384 kbit/s—for pedestrians
  144 kbit/s—for mobile subscribers

Higher transmission speeds provide users of these networks with more services. First, this applies to mobile access to Internet resources at a speed that satisfies the consumer. Mobile telecommunications networks of the SAE/EPS project allow you to transmit and receive large amounts of data, video images, music files and other multimedia information in real time [10]. At the same time, the services of networks of previous generations, which have already proven themselves, should be actively used in the networks of the SAE/EPS project, thereby confirming the fundamental need to implement and expand the concept of VTF. These services, within the concept of VTF, first of all, need to include the service of short message exchange SMS (Short Message Service). An alternative to the already proven SMS and EMS (Enhanced Message Service) is the multimedia message service MMS (Multimedia Message Service). As a conclusion, it can be noted that in accordance with the standards of ITU-T, ETSI and 3GPP in the mobile telecommunications networks of the SAE/EPS project, it is advisable to use the VTF concept to enable its subscribers to use the services of the home network while being outside it—in the business network. The main task of each SAE/EPS network operator should remain the task of providing any subscriber with the usual set of services while roaming, that is, to provide 100% support for the usual functions, despite the fact that such functions will be "virtual", that is – unusual for a business network. The number and variety of services that can potentially be implemented using such systems are very large, which is attractive for both subscribers and mobile operators.

As can be seen from Fig. 2, the functional elements can be physically combined or distributed over the network. It all depends on the features of the demanded products used and the network itself. For example, "Anchor" 3GPP is allowed to be placed together with the subscriber management module, although this is not a requirement. In addition, MME and UPE modules can be combined or located in different network nodes.

The fundamental difference between the fourth-generation mobile telecommunications networks and the previous-generation mobile networks is the ability to provide a higher speed of information transmission. The continuation of the logical development of high speeds will be the application of standards of the fifth, sixth, and seventh generations. Many companies have already announced modems for such networks that use additional high-speed channel capabilities, which, theoretically, are 326.4 Mbit/s for "download", and 172.8 Mbit/s for return.

As an example, according to the ITU-T guidelines, the 5G network speed should have a peak data rate of 20 Gbit/s for the downlink and 10 Gbit/s for the uplink. Latency in a 5G network can be up to 4 ms in a mobile scenario, and can be up to 1 ms in Ultra Reliable Low Latency communication scenarios. Not only will physical users be connected to each other, but also technological devices (cars, automobiles, city infrastructure, public security, and more) will be able to interact with each other [11, 21]. The speed in the standards of higher generations will be higher several times compared to the standards of lower generations. Now that 7G mobile phone technology is coming, which will be two or three times faster than the 4G standard, there is no doubt that we will need to model and implement new MS technologies.

Higher transmission rates provide more services to users of these networks. First, it concerns mobile access to Internet resources at a speed that satisfies the consumer. SAE/EPS mobile telecommunications networks allow you to transfer and receive large amounts of data, video images, music files and other multimedia information in real time. At the same time, the services of previous generations networks, which have already proven themselves, should be actively used in the networks of the SAE/EPS project. Thereby confirming the fundamental need for the introduction and expansion of the VTF concept. These services, within the framework of the VTF concept, first, should include the service of short SMS (Short

Message Service). As an alternative to the already proven services of SMS and EMS (Enhanced Message Service) came multimedia messaging service (MMS). However, when different standards interact, problems arise (and will continue to arise in the future) to optimize the transition from one standard to another, when users are in different coverage planes, with different standards.

Services provided by modern contact centres (call centres) have become very popular in mobile telecommunications. The introduction of contact centres is one of the many ways to organize effectively interaction with their customers when they have any questions or problems: this is their main function in the VTF concept implementation.

The main task of each SAE/EPS network operator is to provide any subscriber with the above-described usual set of services when he is roaming, that is, to provide 100% support for the usual functions despite the fact that such functions will be "virtual", that is unusual for a visiting network. The physical level of the draft standards for "higher generation" mobile communication systems should provide not only higher data rates, but also better spectral efficiency compared to previous standards, frequency compatibility when switching MS during its operation in different regions, which may have equipment with different characteristics.

As an example, the development of fifth-generation technologies provides a certain set of "input" characteristics that serve as a guide for the new standard. For example, compared to the best existing LTE networks, the data transfer rate in 5G networks should be 10–100 times higher, the response time—5 times less, the network should support 100 times more devices. It is assumed that the sequential transition from fourth-generation to fifth-generation networks will take longer than the transition from 3G to LTE.

The number of devices that are connected to the Internet and among themselves is constantly increasing. There is a need for better networks that can ensure the interaction of various standards of mobile communication systems, taking into account their further prospects for improvement. The new generation of networks opens up new opportunities in many areas—from improving the efficiency of production processes, improving safety on the roads and in the city as a whole to improving public services and a cleaner environment.

The mass appearance of 5G solutions is unlikely to take place before 2020–2022, because Telecom operators need to optimize the cost of advanced technologies and related equipment, and determine the priority business tasks in a new area for them. Today, TM operators must also provide for the interaction of "higher" standards that will be developed and are being developed with "lower" standards that have already been implemented and are in operation.

According to experts from telecommunications networks, in order to successfully implement systems of different generations for a number of applications, the frequencies of "higher" generation standards with a low-frequency part of the spectrum (below 6 GHz) will be needed simultaneously.

It is expected that low frequencies (e.g., 3.4–3.6 GHz) and wide bands will allow mass availability of speeds up to 100 Mbit.

Low frequencies and good coverage will ensure good permeability in the premises of various technologies, which is important in order to achieve mass-market expansion. This applies primarily to the 700 MHz band and, to a lesser extent, the 3.4–3.8 GHz band. This range can be used for applications in mobile devices to connect to "higher" networks. As noted above, when implementing the concept of VTF, it is advisable to organize a centralized subscriber database for a particular network, which takes part in the process of using the corresponding virtual technical function. In this case, each switching node of mobile stations must have subscriber data MS, which now are in the zone of its service. When moving a mobile station from the service area of one MSC to the service area of another MSC, they will receive on request the necessary information from the centralized subscriber database, which must have ultra-high power and the ability to interact with the centralized subscriber databases of the global network. However, this issue requires a separate study and study. The number and variety of services that can potentially be implemented using such systems is very large, which is very attractive for both mobile telecommunications subscribers and mobile operators. It should be noted that the list of services provided is growing rapidly and steadily. Some services may be in demand, which at first glance will not have an economic effect, but over time, when actively served in advertising brochures, such a service may acquire the effect of "explosion", which will provide rapid coverage of the expensive costs of upgrading the next standard. As it was already noted earlier, one or more service devices (components) of both the business and home (supporting) network of subscribers can participate in the implementation of a certain service. The complexity of obtaining is due, first, to the variety of options for building alarm networks in which applications are transmitted between these OP, and the large volume of alarm systems used on these networks. Today, the main alarm system in cellular communication networks is the ACS system No. 7, which according to ITU-T recommendations [3, 11] can be used in mobile telecommunications networks of the SAE/EPS project. At the same time, the proposed approach to determining the value in the network of ACS No. 7 can be used in any other packet-signalling network, built based on a family of other protocols may have different concepts with different components, especially in the protocols of mobile communication standards of "higher generations".

4 Results

Currently, there are two concepts providing additional services in mobile telecommunications networks:

1. Nodes of the services (Service nodes)
2. "Classical" intelligent communication network IN

The first one is used for the above-mentioned services organization: SMS, EMS, MMS, contact centre services, etc. Main value of the second concept for mobile telecommunications networks is the lists of services capability set (CS). So far, the main idea is to separate the processes of traditional switching from the procedures for the new services provision. The relevance of this idea makes it attractive for mobile telecommunications networks of the SAE/EPS project, focused on providing additional services to the subscriber.

The architecture of IN and the mobile telecommunications networks are very similar (Fig. 3). However, mobile telecommunications networks are not able to provide adequately the independence principle from services inherent in the concept of IN [1, 2, 5–10, 13, 14]. Given the above, the mobile telecommunications operators seek to seize the advantages offered by the concept IN. The implementation of the same service in IN occurs through a certain sequence functions execution. Under the function, it should be understood the mobile telecommunications network set of action components in the process of services implementation at a single address to the application. At the same time, during the service implementation, each component can perform one or more functions. One of the main advantages offered by mobile telecommunications networks is free subscriber possibility movement from its zone to one or another zones served by other mobile telecommunications networks operators. At the same time, the subscriber would like to keep his personal services set (profile) in other networks (to be able to roaming services). However, unlike wired networks, in mobile telecommunications switches, trigger points and

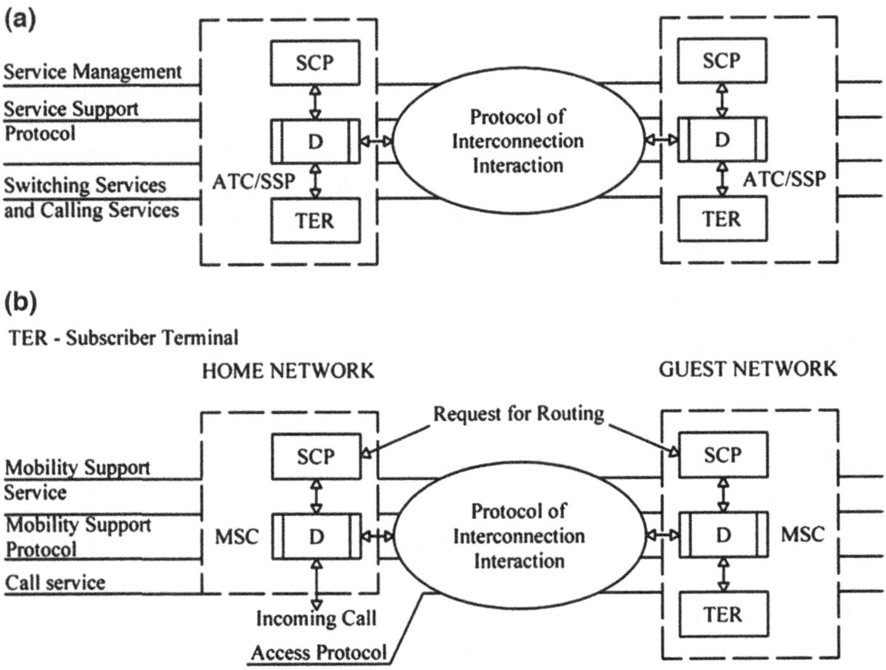

Fig. 3 Architecture of IN (**a**) and mobile telecommunications networks (**b**). (Source: [14])

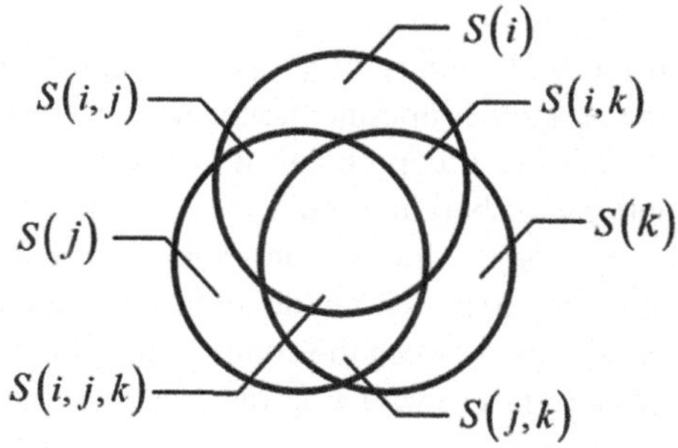

Fig. 4 Model of service coverage in project networks SAE/EPS. (Source: [14])

service profiles are not represented as static data, but are defined at registration. The possibilities of home and visiting networks can be different, which will affect the list and characteristics of services that can be offered to the user that has moved to the visiting network [13, 14].

For the Fig. 4 abstract model with three mobile telecommunications of the fourth generation systems belonging to different operators is shown.

Each of the three SAE/EPS project systems shown in Fig. 4 has its own set of services: $S(i)$, $S(j)$, $S(k)$ and are shown by the circles. The intersections of these circles are services that are the same for these networks. As the number of SAE/EPS project systems increases, the overall platform $S(i, j, k)$ decreases. According to the standards ITU-T ETS1, 3GPP in the mobile telecommunications networks of the SAE/EPS project, to enable its subscribers to use the functions of the home network while outside it, that is, in the visiting network, the previously mentioned VHS concept was introduced, the list of functions of which was quite limited and had a declarative character. The degree to which the proposed VTF concept can meet the actual needs of subscribers will be extremely high and will depend only on the degree of cooperation between operators, their technical capabilities, compatibility of user equipment, etc. In addition, the SAE/EPS project systems twill offer their subscribers a global roaming service should support VTF, if these systems have the capabilities of a subscriber service similar to the VHS service from the very beginning.

The physical layer of the SAE/EPS project systems provides not only higher data rates, but also better spectral efficiency compared to the "lower" standards. As already noted, the interaction of various mobile communication standards will require the development and implementation of appropriate equipment.

In the process of implementing the VTF concept, a certain number of services are performed using hardware and software. As mentioned earlier, in the process of implementing the service, each component can perform one or more functions. The composition of functions depends on many factors, for example, on the service itself, the actions of subscribers, etc. The resulting functions can be located in various components of both the home (support) and visiting network. In this case, the various functions can be located in the same components. Thus, there is a set of variants (scenarios) of the VTF structurally functional construction concept organization. Since there are various options for structurally functional construction of the VTF concept, then at the stage of its design there is a problem of choosing one or another option – a scenario for the concept implementation.

To solve this problem, it is necessary to have a tool that allows evaluating the options for building the VTF concept. However, as noted in scientific publications on this topic, often are only the scenarios for the implementation of such concepts and their analysis is not done. In some other works, the volume of signal information circulating between the components of mobile telecommunications networks acts as a criterion for evaluating the construction options. At the same time, less attention is paid to assessing the impact on the quality of services scenario choice, and hence on the degree of these end-user services satisfaction [13–15, 21]. In this case, there is a need for almost simultaneous interaction of different components in different mobile communication standards.

5 Conclusions

The state and prospects of mobile communication systems development are considered. The definition of concepts, individual scenarios implementation is given. An abstract model with three fourth-generation mobile communication systems belonging to different operators is shown. Considering the results of preliminary studies, it is concluded that in the concepts implementation, it is advisable to organize a centralized subscriber database for a specific network, which takes part in the process of a virtual technical function providing.

The general picture of the next generation's mobile communication standards harmonization is formed. The general purpose of the study and partial tasks that need to be addressed are the further concept developments of telecommunication networks of mobile communication (fifth generation).

The improvement of virtual technical functions will ensure a strong market position in telecommunication systems.

The latest network technologies are a natural product of scientific and technological progress, which is rapidly being carried out by the technically developed countries of the world to obtain the advantages of leadership in the global distribution of labour. Ukraine is forced to play the role of an importer of the latest network technologies, paying for their import with other goods or services of its own production. This fee will be greater the more complex and more ready-to-use are the means are in which network technologies are implemented. The most effective technologies that should be used for the development of info communication in Ukraine should be considered further research of mobile communication standards of the latest and future generations, research, testing and evaluation of a variety of new technologies that determine the development of telecommunications in the world.

Acknowledgements This paper has been written within VEGA project Nr. 2/0077/19: Work competencies in the context of Industry 4.0 development and KEGA project Nr. 030STU-4/2018: E-platform for improving collaboration among universities and industrial enterprises in the area of education.

References

1. FY 1998 National Communications System, 1998, http://www.ncs.gov/library/reports/ncs_fy1998.pdf
2. ITU-T Recommendation F.115, Service objectives and principles, 1995
3. P. Bakhovskyi, M. Yevsiuk, Specific aspects of the development of mobile telecommunications network. Perspective Technologies and Devices, Issue 13 (2018)
4. M. Torabi, R. Buhrke, Third-generation mobile telecommunications and virtual home environment: A prioritization analysis. Bell Labs Techn. J. **3**(3), 50–62 (1998)
5. D. MacDonald, Teletraffic performance in intelligent network services. ITC-14, 1994
6. R. Andrade, L. Logrippo, MoRaR: A Pattern Language for Mobility and Radio Resource Management, 2004, http://www.site.uottawa.ca/~luigi/papers/06_andrade_logrippo.pdf
7. ITU-T Recommendation Q.I751, Internetwork signalling requirements for IMT-2000 capability set 1, 2000
8. ITU-T Recommendation Q.I741.1, IMT-2000 references to release 1999 of GSM evolved UMTS core network with UTRAN access network, 2002
9. ITU-T Recommendation Q.I741.2, IMT-2000 references to release 4 of GSM evolved UMTS core network with UTRAN access network, 2002
10. ITU-T Recommendation M.816-1, Framework for services supported on International Mobile Telecommunications-2000 (IMT-2000), 1997
11. Electronic resource, https://www.sdxcentral.com/5g/definitions/imt-2020, 2019
12. ITU-T Recommendation Q.1702, Long-term vision of network aspects for systems beyond IMT-2000, 2002
13. P. Bakhovsky, Intellectual network: Socio-economic concept, in *Conference Modern Problems of Telecommunications 2008*, Lviv (2008), pp. 65–68
14. P. Bakhovsky, *Materials of the Dissertation Are to Obtain a Scientific Degree of the Candidate of Technical Sciences* (Ukrainian State Academy of Railway Transport, Kharkiv, 2010)
15. M. Bawa, D. Cagáňová, Selecting network protocols for internet of things based upon innovation and knowledge management. J. Telecommun. Syst. Manag. **7**(2), 1–4 (2018)
16. P.F. Bakhovskyi, E.D. Amirkhanov, Y.I. Toroshanko, K.V. Khmara, Analysis of the economic feasibilities of introduction of technologies of VTF in the networks of SAE/EPS. Scientific Notes of the Ukrainian Research Institute of Communication, vol. 1(41) (2016)
17. P.F. Bakhovskyi, V.O. Hrebennikov, Trends in the development of the telecommunications market in Ukraine and proposals for its improvement. Scientific Notes of the Ukrainian Research Institute of Communications, vol. 3(5) (2008)
18. 3G Evolution Horizons: Architectural Aspects, http://osp.ru/nets/2007/16/4626460/
19. Y. Toroshanko, Y. Selepyna, N. Yakymchuk, V. Cherevyk, Control of traffic streams with the multi-rate token bucket, in *International Conference on Advanced Information and Communications Technologies, AICT 2019* (2019), pp. 352–355
20. M.A. Vinohradov, P.F. Bakhovskyi, Investigation of a non-stationary random access network with a dynamic protocol under heavy load conditions in the framework of the VTF concept implementation. Scientific Notes of the Ukrainian Research Institute of Communications, vol. 4(12) (2009)
21. D. Cagáňová, M. Bawa, D.R. Delgado Sobrino, A. Saniuk, *Internet of Things and Smart City*, 1st edn. (Uniwersytet Zielonogórski, Zielona Góra, 2017), p. 138

IoT-Based Household Energy Consumption Prediction Using Machine Learning

Pouya Jafarzadeh, Fahimeh Farahnakian, Jari-Pekka Paalassalo, and Osmo Eerola

1 Introduction

Digitalization, urbanization, and the reduction of non-renewable energy resources raise concerns on the availability of energy in the future. Increased energy consumption would increase the need for fossil fuels that cause irreparable harm to our environment. According to the report issued by the International Energy Agency Organisation for Economic Co-operation and Development (OECD), energy consumption in the building sector accounts for 40% of total consumption and 36% of CO2 emissions in the EU [1]. In order to ensure the availability of energy and to preserve our environment in the future, we need to find more efficient and intelligent solutions for reducing energy consumption in the building sector. In addition, these solutions can help building managers to make better decisions so as to reasonably control all kinds of equipment

Developing smart houses are an efficient way for energy saving. A smart house is a set of many sensors, controllers, relays and meters, as well as systems for controlling them. The purpose of a smart house is to collect useful data about residents' behaviour, habits and environment in order to make decisions efficiently for reducing energy usage. With machine learning techniques, a smart house can learn residents' lifestyles and routines, and thus automates operations. For example, light and temperature can be automatically adjusted in the presence and absence

P. Jafarzadeh (✉) · J.-P. Paalassalo · O. Eerola
Faculty of Engineering and Business, Turku University of Applied Sciences, Turku, Finland
e-mail: pouya.Jafarzadeh@edu.turkuamk.fi; jari-pekka.paalassalo@turkuamk.fi; osmo.eerola@turkuamk.fi

F. Farahnakian
Department of Future Technologies, University of Turku, Turku, Finland
e-mail: fahfar@utu.fi

D. Cagáňová et al. (eds.), *Advances in Industrial Internet of Things, Engineering and Management*, EAI/Springer Innovations in Communication and Computing,
https://doi.org/10.1007/978-3-030-69705-1_8

of residents. Moreover, smart home systems can be easily integrated with existing conventional systems and switchboards to optimize the energy consumption of the house [2]. Most smart houses employ the Internet of Things (IoT) for connecting physical and virtual devices in a house to the Internet. IoT allows users to remotely manage, collect data and automate the operation of their device or system. IoT can be utilized in the small scale and simple appliances, such as a refrigerator in a house and a washing machine, or as large as the transport infrastructure of a city.

Energy consumption prediction of a household can very significantly save energy and reduce environmental effect [3–5]. Therefore, tremendous efforts have been invested in energy optimization solutions for smart homes. The most important of these solutions is providing minimum energy consumption while maintaining customer comfort [6]. For this purpose, lots of machine learning (ML) algorithms have been proposed for energy prediction of appliances in smart homes [7, 8]. Besides the conventional ML methods like neural networks, multi-layer perceptron, k-nearest neighbor, super vector machines, decision trees, the deep learning, hybrid and ensemble methods have been significantly increased.

In this chapter, we present a framework for energy consumption prediction of a house using machine learning methods. The chapter also addressees the changes in factors such as the presence and absence of residents, as well as environmental impacts such as temperature and humidity, and the effect of ambient luminous intensity on energy consumption. To collect the real data, we used the Arduino micro-controller board and different type of sensors which are installed on the board. The data obtained from the sensors were analysed to determine the correlations between the possible factors influencing energy consumption. The energy consumption prediction was performed using three popular machine learning methods: multi-layer perceptron (MLP) and K-nearest neighbour (K-NN) and long short-term memory (LSTM). The experimental results on real data show that MLP-based prediction model can perform better than K-NN for energy prediction at the current time. In addition, LSTM can forecast hourly energy consumption with high accuracy.

The remainder of the chapter is organized as follows. Section 2 discusses some of the most important related works. The proposed energy consumption prediction framework is introduced in Sect. 3. Section 4 describes the experimental design and setup. Experimental results are presented in Sect. 5. Finally, the conclusion is presented in Sect. 6.

2 Related Work

Designing efficient energy consumption prediction methods can play an important role in improving energy saving and reducing environmental effects. For this reason, various approaches have been proposed for energy consumption prediction in the recent literature. The majority of these approaches employ the historical energy consumption time series data and build a prediction model using ML algorithms. For instance, TSO et al. [6] have utilized linear regression (LR), decision tree (DT) and artificial neural network (ANN) for prediction of energy consumption in government buildings, dwellings and cottages. In addition, they have investigated the effect of the number of residents, the size of the family, the time of construction, the type of housing and the electronic equipment in use. They collected the results for two phases: summer and winter. Their results show that DT has more accuracy compared to ANN and LR.

Similarly, three models based on simple linear regression (SLR) and linear regression multiple layers (LRML) are presented in [9] for energy consumption prediction. In addition, they considered three different cases. In the first case, the training dataset related to energy consumption was grouped on an hourly basis, in the following case each year and lastly on the basis of daily peak hours. Their results show that the SLR-based algorithm was more able to predict hourly and year-based forecasting compared with the LRML.

Truong et al. [10] proposed to predict energy consumption based on a graphical model. They aim to find the relationship between energy consumption and the human actions and routines, and in particular the utilization rate and time of use of electronic household appliances. Olofsson et al. [7] proposed a neural network model for long-term energy demand prediction based on short-term training data. The model parameters include temperature difference between indoor and outdoor environment and energy required for heating and other utilities.

Fayaz et al. [8] proposed extreme learning machine (ELM), adaptive neuro-fuzzy inference system (ANFIS) and ANN to predict energy consumption in residential buildings. The aim of this study was to forecast energy consumption for the following week and the following month. Moreover, they have employed different numbers and types of membership functions to obtain the optimal structure of ANFIS. For ANN, they have also tested several architectures with different number of hidden layers. Kreider et al. [11] reported results of a recurrent neural network on hourly energy consumption data to predict building heating and cooling energy needs in the future, knowing only the weather and timestamp.

Liet al. [12] utilized back propagation ANN, radial basis function neural network (RBFNN), general regression neural network (GRNN) and support vector machine for predicting the annual electricity consumption of buildings. These models are trained on a real data of 59 buildings and tested on nine buildings. Their results show that GRNN and SVM were more applicable to this problem compared with other models. However, the test conducted to all methods indicated that SVM could predict with higher accuracy and was better than the others.

Samuel et al. [13] addressed the energy consumption prediction problem using support vector machine (SVM) and multi-layer perceptron (MLP). They also improved the forecasting for district consumption by clustering houses based on their consumption profiles. Authors demonstrate an empirical plot of normalized root mean squared error against number of customers and show it decreases. This work aggregates up to 782 homes.

SVM is another ML algorithm that is commonly applied for forecasting energy consumption. SVM is increasingly used in research and industry due to its highly effective model in solving non-linear problems [3]. Besides that, since it can be used to solve non-linear regression estimation problems, SVM can be used to forecast time series. This paper [3] reviews the building electrical energy forecasting method using artificial intelligence (AI) methods such as SVM and ANN. Both methods are widely used in the field of forecasting. Dong et al. [14] applied SVM to predict monthly buildings energy consumption in tropical regions. Their results on 3 years data of electricity consumption of show that SVM has a good performance in prediction.

Jain and Satish [15] employed SVM in clustering-based Short-Term Load Forecasting. The forecasting is performed for the 48 half hourly loads of the next day. The daily average load of each day for all the training patterns and testing patterns is calculated and the patterns are clustered using a threshold value between the daily average load of the testing pattern and the daily average load of the training patterns. Their results obtained from clustering the input patterns and without clustering are presented and the results show that the SVM clustering-based approach is more accurate. For the same type of forecasting, Mohandes [16] showed that the SVM result for forecasting was better compared to the Auto Regressive modelling. The RMSE for testing data for SVM was 0.0215, while the value was 0.0376 for AR. The lower RMSE in the analysis indicated that the forecasting was more accurate.

Rahman et al. [17] employed a deep recurrent neural network in order to predict heating demand for a commercial building in the USA. They investigated the performance of the deep model over a medium- to long-term horizon forecasting. Deep learning is a particularly attractive option for longer-term forecasts as they are able to model expressive functions through multiple layers of abstractions [18].

3 The Proposed Energy Consumption Prediction Framework

In this section, we describe the proposed framework for energy consumption prediction. Figure 1 shows an illustration of our whole framework. The framework consists of four main layers as follows:

3.1 Data Collection Layer

This layer consists of physical IoT devices including various sensors. The sensors are used to collect the contextual information environmental conditions, circumstances, temperature, humidity and user occupancy. These sensors include:

- **Humidity and temperature sensor** provides the indoor temperature real data of the house, as well as the humidity percentage. The humidity and temperature

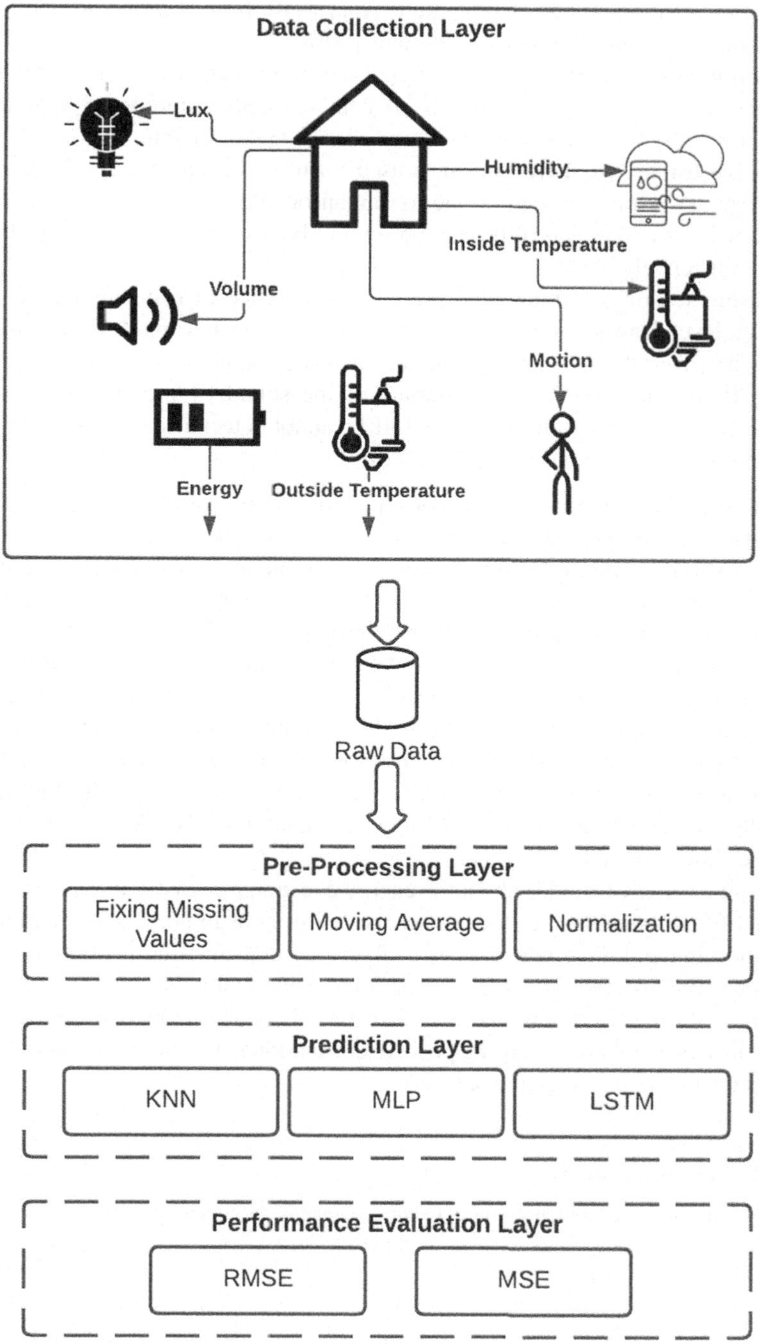

Fig. 1 An overview of the proposed framework for energy consumption prediction

sensor can measure the temperature from −40 °C to 125 °C. In addition, it measures the humidity between 0% and 100%.

- **Illuminance intensity sensor** is a digital sensor that can measure the illuminance value between 0.1 and 40.00 Lux. It contains both infrared and visible light spectrum diodes, which help to obtain accurate measurement.
- **Passive Infra-Red (PIR) sensor** is used for user occupancy detection by rapid changes in infrared radiation due to human or other influence movement. PIR sensor information is 1 or 0 for representing busy or nor busy state for the past 1 h, respectively.
- **Volume sensor** determines the presence or absence of residents based on the sound level. This sensor is based on a simple microphone and LM386 amplifier, which convert the surrounding sound to an analog signal with sensitivity of about 50 dB. It can measure the amplitude of the sound by measuring the analog voltage. The value of this sensor is 0 if it cannot detect any sounds at the house in the past 1 h, otherwise is 1.

Data was collected from this sensor set continuously from November to December 2018. All sensors' measurements were recorded every 5 s. All selected sensors measure variables that have direct influence on energy consumption. The energy consumption that is obtained by the local electricity utility represents the total value of energy consumption (kilowatt-hour (kWh)) for each hour. For this reason, we consider the average value of lux, temperature, humidity for 1 h. In addition, the maximum value of PIR and volume within 1 h is assumed. Moreover, we got the outdoor temperature data (°C) from the local weather bureau which is located in 6 km far from the house. Furthermore, the input variables used from our data include illuminance, indoor temperature, outdoor temperature, humidity, PIR and volume while the output variable is the total energy consumption. The characteristics of the data are shown in Table 1.

Figure 2 can give an idea about the energy consumption distribution of the house per hour for an example day. You can see the energy consumption is low in midnight, early morning and noon whereas the peak energy consumption is observed during afternoon timings. This is due to the fact that residential buildings have maximum occupancy during afternoon times which results in higher energy consumption. This also indicates that there is a strong positive correlation between energy consumption and occupancy in the collected dataset.

Table 1 The data characteristics

	Illuminance	Indoor temperature	Outdoor temperature	Humidity	PIR	Volume	Energy
Mean	51.87	20.69	−1.72	25.87	0.52	0.44	2.83
Std	118.30	0.47	3.11	3.07	0.50	0.49	1.71
Min	1.00	19.88	−8.00	21.06	0.00	0.00	0.99
Max	821.47	22.51	3.50	39. 15	1.00	1.00	13.29

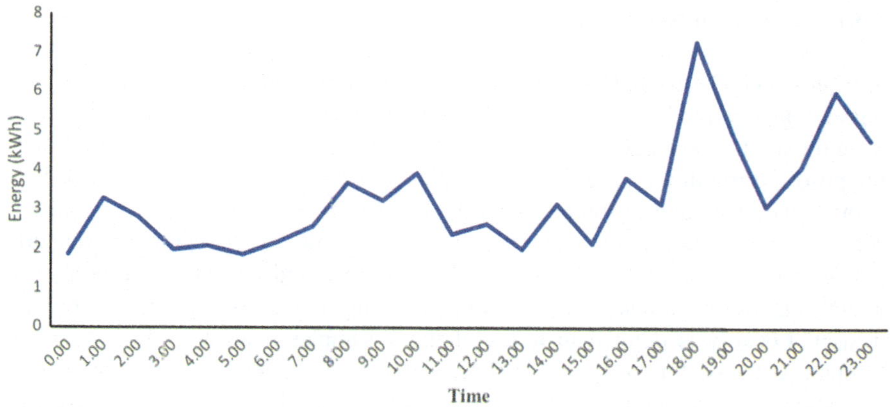

Fig. 2 Hourly distribution of energy consumption data for 28 November

3.2 *Pre-processing Layer*

In this layer, we clean the data by removing outliers and fixing the missing values in order to prepare the data in the format required by the upper layer. In addition, we applied data normalization to generate the same range of values for each measurement for using machine learning models. Therefore, the sample data was normalized to fit into the range between 0 and 1 by using the following equation:

$$x_n = \frac{x_i - x_{min}}{x_{max} - x_{min}}, i = 1, 2, \ldots, n \tag{1}$$

where x_n is the normalized values, x_i is i of input data. x_{max} and x_{min} represent the maximum and minimum values of data x_i, respectively [8]. n is the number of samples.

3.3 *Prediction Layer*

In the prediction layer, we utilized two well-known machine learning algorithms (K-NN and MLP) for energy consumption prediction at the current time. Moreover, we used LSTM algorithm for 1 h energy consumption prediction for the residential house. Here, we briefly review the machine learning methods which were used in this work for energy consumption prediction.

3.3.1 K-Nearest Neighbour

K-nearest neighbour (K-NN) [19] is one of the most widely used algorithms for regression supported by its simplicity and intuitiveness in finding similar instances in multivariate and large-dimensional feature spaces of arbitrary attribute scales. Suppose the training dataset has m samples that each sample x_i is described by n input variables and an output variable y_i such as $x_i = \{x_{i1}, x_{i2}, \ldots, x_{in}, y_i\}$. The goal is to learn a function $f : x \rightarrow y$ known as a regression function that model the relationship between input variables and an output variable. The K-NN regression estimates the function by taking a local average of the training dataset. Locality is defined in terms of the k samples nearest to the estimation sample. As the performance of K-NN algorithm strongly depends on the parameter k, finding the best values of k is essential. A large k value decreases the effect of noise and minimizes the prediction losses. However, a small k value allows simple implementation and efficient queries [20, 21].

Cross validation can employ to estimate the accuracy and validity of the classifier with different k values [21]. Two common schemes of cross validation are k-fold cross validation and leave-one-out cross-validation (LOOCV). In k-fold cross-validation, the dataset is randomly divided into k subsets or folds and repeated k times. Each time, one fold is reserved as a test dataset for validating the model and the remaining $k - 1$ folds are used for training the model as a training dataset. Then, the classification accuracy across all k trails is computed. LOOCV is a particular type of k-fold cross-validation where k is the number of examples in the dataset. As the size of the dataset is not large, LOOCV is utilized in this paper.

3.3.2 Multilayer Perceptron

Multilayer perceptron (MLP) [22] is an artificial neural network (ANN) for modelling non-linear relationship between input and output variables through multiple layers of interconnected processing elements. ANNs are inspired from the human brain to predict the new observation form the previous observation after executing a learning process. MLP consists of three main layers: input, hidden and output layers. Each interconnection between neurons has associated with a scalar weight. The MLP model is trained by adjusting weights in order to minimize the error. The error in the final layer can be calculated as follows:

$$E = (y_i - \hat{y}_i)^2 \tag{2}$$

where y and $\hat{y}$ are the actual or measured output and the predicted output, respectively.

During training the model, the weights of each node j at layer l are updated using stochastic gradient descent as follows:

$$w_{ji}{}^{m} = w_{ji}{}^{m} - \alpha(\frac{\partial e}{\partial w_{ji}{}^{m}}) \tag{3}$$

where α is the learning rate. The partial derivatives $(\frac{\partial e}{\partial w_{ji}{}^{m}})$ are determined by back-propagation method [18].

However, a gradient descent learning approach poses two problems [23]. First problem is overfitting. It means MLP can adjust weights for performing well on the training dataset but it is not able to produce the accurate response for a new data. This problem can be solved by splitting the training dataset into two parts: one part for training and the part for validation. Then, the algorithms is stopped when the error increases on the validation dataset.

The second problem is local minima for finding the global optimal solution by exploring the search space. One way for avoiding local minima is using adaptive learning rate. An adaptive learning rate dynamically changes the gradient descent step size, such that the step size is larger when the gradient is steep and smaller when the gradient is flat.

3.3.3 Long Short-Term Memory

Long short-term memory (LSTM) [24] networks are a type of recurrent neural networks (RNNs). As RNNs have feedback connections unlike MLP, they are most common to modelling sequences and temporal dependencies. An activation function in LSTM accounts the temporal dependency by considering the previous timestep $(t-1)$. LSTM consists of input gate i, forget gate f, output gate o and cell state c. Generally, the three gates control the information flow.

At current time t, the input is x_t, the hidden layer output is h_t and its former output is h_{t-1}, the cell input state is $\hat{c}_t$, the cell output state is c_t and its former state is c_{t-1}. For measuring c_t and h_t, firstly the following equations are calculated in order:

$$i_t = \sigma(w_1^i.x_t + w_h^i.h_{t-1} + b_i), \tag{4}$$

$$f_t = \sigma(w_1^f.x_t + w_h^f.h_{t-1} + b_f), \tag{5}$$

$$o_t = \sigma(w_1^o.x_t + w_h^o.h_{t-1} + b_o), \tag{6}$$

$$\hat{c}_t = \tanh(w_1^c.x_t + w_h^c.h_{t-1} + b_c), \tag{7}$$

where w_1^i, w_1^f, w_1^o and w_1^c are the weights matrices connecting x_t to the three gates and the cell input. w_h^i, w_h^f, w_h^o and w_h^c are the weights matrices connecting h_{t-1} to the three gates and the cell input. b_i, b_f, b_o and b_c are the bias terms. σ and tanh represent the sigmoid and tangent functions, respectively.
Secondly, the cell output state is measured as follows:

$$c_t = i_t \times \hat{c}_t + f_t \times c_{t-1}, \tag{8}$$

Finally, the hidden layer output is calculated as:

$$h_t = o_t \times \tanh(c_t) \tag{9}$$

3.4 Performance Evaluation Layer

The final layer, the performance evaluation layer, evaluates the performance of the proposed models based on two common performance metrics: mean square error (MSE) and root mean square error (RMSE). They are calculated with the following equations:

$$MSE = \frac{1}{n}\sum_{i=1}^{n}(y_i - \hat{y}_i)^2 \tag{10}$$

$$RMSE = \sqrt{\frac{1}{n}\sum_{i=1}^{n}(y_i - \hat{y}_i)^2} \tag{11}$$

where y is the real energy consumption, $\hat{y}$ is the predicted energy consumption and n is the number of samples.

4 Experimental Setup

We collected a real dataset using a sensor system in a family house that is located in southwest of Finland and was constructed in the year 2018. This house has a structure which is typical for a country such as Finland. The major construction material is concrete with wood and glass windows. The house has an exhaust air heat pump (EAHP) and is heated by underfloor heating. Typically, the house heating is a closed loop control system that is controlled either by inside or outside temperatures. The control system tries to keep the room temperature constant at the preset value by residences (20–22 °C).

Energy consumption data is recorded by the electrical utility on the hourly interval. The three major energy consuming systems are: heating, lighting and electrical equipment. For the experiments, we proposed 80% of the whole data for training the models and 20% for testing.

For the hardware implementation, we used Arduino mega micro-controller as it is inexpensive, cross-platform, simple, clear programming environment, open-source and has sufficient digital/analog ports and extensible software support [25].

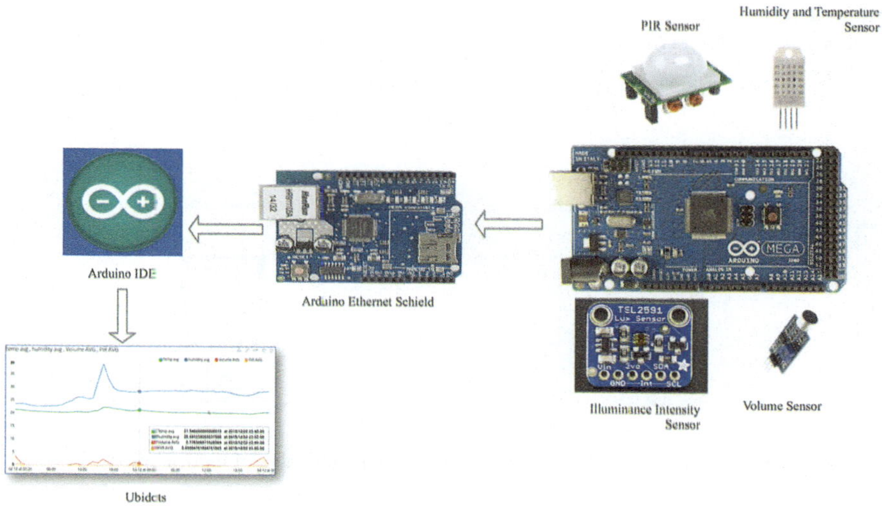

Fig. 3 An overview of the hardware and software implementations

This micro-controller has 54 digital pins and 16 analog pins. Figure 3 shows an overview of the hardware and software implementations. Several sensors are used to monitor the smart house: the temperature sensor, humidity sensor, illuminance intensity sensor, PIR motion sensor and volume sensor. All sensors are connected to the pin of Arduino. In order to connect Arduino to the Internet for IoT development, we connected Arduino to the Ethernet shield. The Ethernet shield operating voltage is 5 V (supplied from the Arduino Board) and the connection speed is 10/100 Mb. For Arduino programming, we utilized an environment that is called Arduino IDE. Arduino IDE supports C and C++ programming languages. We used the Ubidots platform for collecting the sensor's output in Cloud.

5 Experimental Results

We evaluated two popular ML methods (K-NN and MLP) for the energy consumption prediction at the current time. In addition, LSTM is used for predicting the hourly energy consumption.

The obtained results show that K-NN-based prediction model can get very high accuracy for the test dataset since the RMSE and MSE values are 1.80 and 3.25, respectively. To obtain the best K value, we utilized the leave-one-out cross-validation (LOOCV). We assumed the value of K from 1 to 10 and got the maximum accuracy when $K = 2$.

Table 2 The number of neurons of the hidden layer in each model

Model	Layer1	Layer2	Layer3
MLP (1 hidden layer)	6	–	–
MLP (2 hidden layers)	6	6	–
MLP (3 hidden layers)	6	6	6

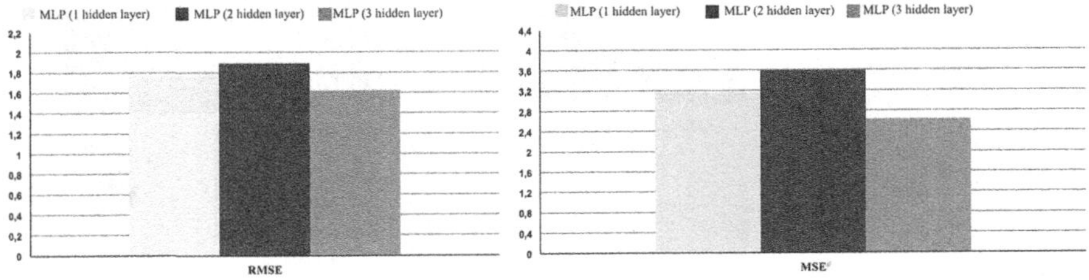

Fig. 4 RMSE and MSE of MLP architectures with different number of hidden layers

As the number of neurons and layers can significantly effect on MLP performance, we test three different MLP models with one, two and three hidden layers. Table 2 shows the number of neurons of the hidden layer in each model.

Adam optimization [26] is used to train the model with the learning rate 0.00001 and reduction factor of 0.1. The epoch number was 20 and the number of iteration in each epoch was 1000. As you can see in Fig. 4, MLP-based model with three hidden layers can predict the energy consumption with the minimum RMSE (1.62) and MSE (2.63). Generally, the experimental results show that an MLP-based model is better in prediction than the K-NN-based model. As the aim of this study is to develop accurate prediction model, MLP has been selected. Moreover, Fig. 5 illustrates the prediction results of four proposed models in two training and test phases. The results show that the MLP-based models perform better than K-NN. This supports the claim that the MLP can adequately account the relationship between the measured input variables and output (energy consumption).

In addition, we evaluate our framework based on LSTM for forecasting the energy consumption for the next 1-h period. We proposed three LSTM layers which each layer has 30 neurons. Moreover, the value of batch size and epochs are 32 and 200, respectively. In addition, the best optimizer for our neural network model in order to learn properly and tune the internal parameter is Adam based on our experiments. Figure 6 demonstrates the predicted energy consumption obtained from LSTM model in comparison with the actual consumption for test dataset.

The results show that both lines almost coincide, which indicates that our prediction algorithm is working nicely. MSE and RMSE for the LSTM model is 0.0062 and 0.0789, respectively.

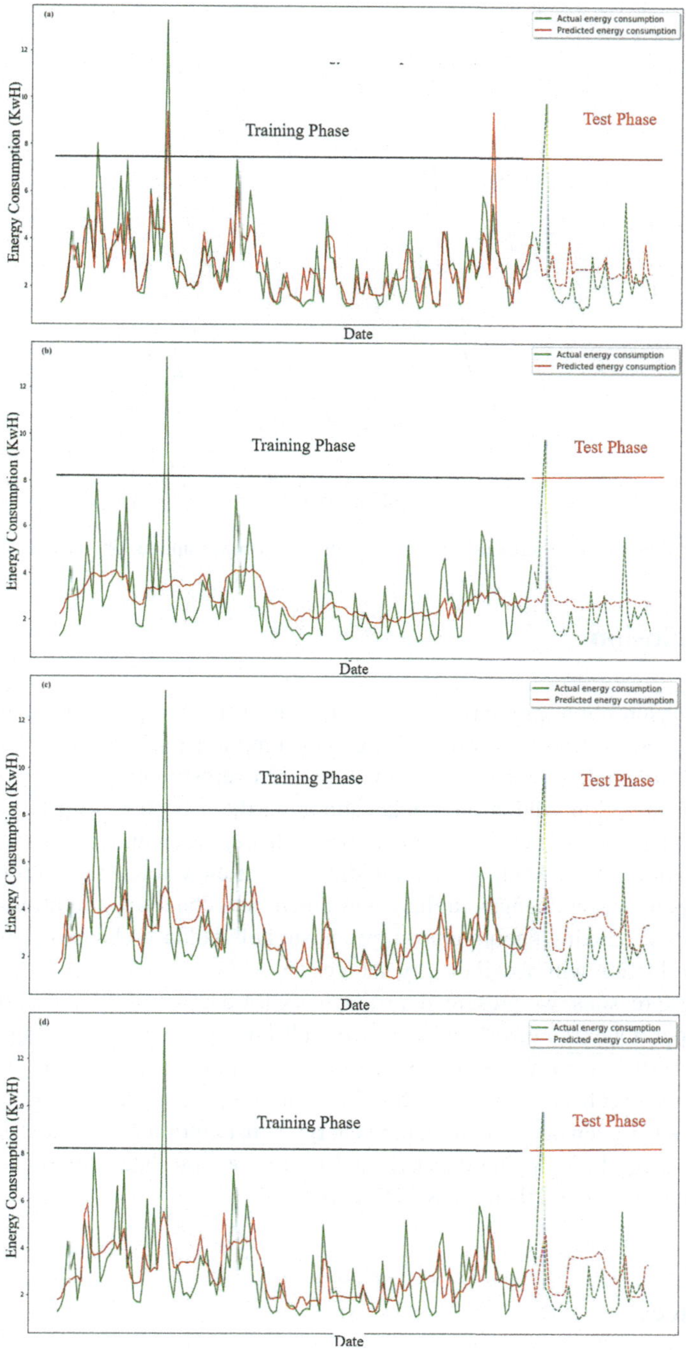

Fig. 5 The prediction results of (**a**) K-NN, (**b**) MLP with one hidden layer, (**c**) MLP with two hidden layers and (**d**) MLP with three hidden layers between 23 November 2018 and 10 December 2018 in two phases: training and test

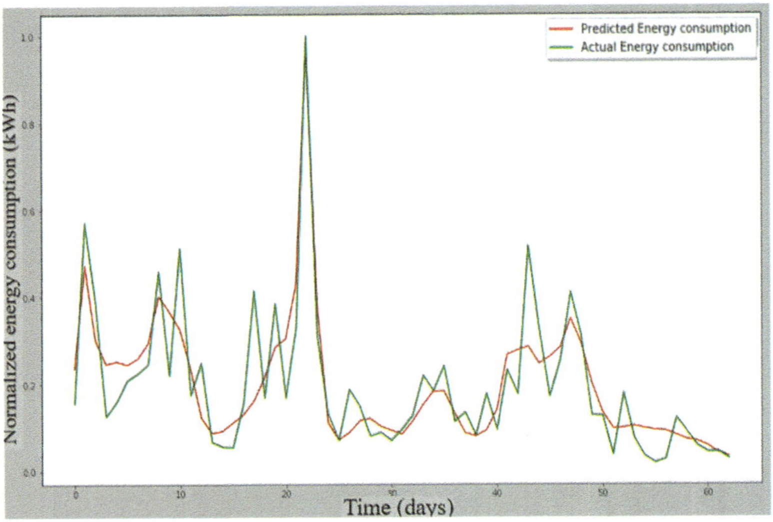

Fig. 6 LSTM prediction results in comparison with actual consumption for test dataset

6 Conclusion

Energy consumption prediction is an efficient way for planning and balancing energy production according to user consumption requirements. The energy provider can handle energy underestimating and overestimating challenges by the energy consumption prediction. Underestimating the need for electricity will lead to potential power outages and an increase in delivery number of costs while overestimation, would lead to potential idle losses. Renewable energy production is highly dependent on changes in the environment and changing conditions, such as the climate. Predicting energy consumption can help balance changes in renewable energy production with fossil energy sources.

For these reasons, we presented a framework for energy consumption prediction in this paper. Our framework utilized two well-known machine learning methods, K-NN and MLP, for predicting the energy consumption at the current time based on sensor data such as temperature, humidity, motion, sound and user occupancy. In addition, our framework can predict the energy consumption for the next 1-h period according using LSTM. The experimental results on real data show that the MLP model is more accurate than the K-NN model.

References

1. REO, Renewable energy outlook. http://www.worldenergyoutlook.org/media/weowebsite/2012/WEO2012_Renewables.pdf (2012)

2. M. Soliman, T. Abiodun, T. Hamouda, J. Zhou, C. Lung, Smart home: integrating internet of things with web services and cloud computing, in *2013 IEEE 5th International Conference on Cloud Computing Technology and Science*, vol. 2 (2013), pp. 317–320
3. L. Pérez-Lombard, J. Ortiz, C. Pout, A review on buildings energy consumption information. Energy Build **40**, 394–398 (2008)
4. N. Fumo, A review on the basics of building energy estimation. Renew. Sust. Energy Rev. **31**(C), 53–60 (2014)
5. H.-X. Zhao, F. Magoulès, A review on the prediction of building energy consumption. Renew. Sust. Energy Rev. **16**, 3586–3592 (2012)
6. G. Tso, K. K. W. Yau, Predicting electricity energy consumption: a comparison of regression analysis, decision tree and neural networks. Energy **32**, 1761–1768 (2007)
7. T. Olofsson, S. Andersson, Long-term energy demand predictions based on short-term measured data. Energy Build. **33**(2), 85–91 (2001)
8. M. Fayaz, D. Kim, A prediction methodology of energy consumption based on deep extreme learning machine and comparative analysis in residential buildings. Electronics **7**(10), 222 (2018)
9. C.J. Wallnerström, J. Setréus, P. Hilber, F. Tong, L. Bertling Tjernberg, Model of capacity demand under uncertain weather (2010), pp. 314–318. https://doi.org/10.1109/PMAPS.2010.5528841
10. N.C. Truong, J. McInerney, L. Tran-Thanh, E. Costanza, S.D. Ramchurn, Forecasting multi-appliance usage for smart home energy management, in *23rd International Joint Conference on Artificial Intelligence (IJCAI 2013) (09/08/13)* (2013, April)
11. J.F. Kreider, D. Claridge, P.S. Curtiss, R.H. Dodier, J.S. Haberl, M. Krarti, Building energy use prediction and system identification using recurrent neural networks. J. Sol. Energy Eng. **117**, 161 (1995)
12. Q. Li, P. Ren, Q. Meng, Prediction model of annual energy consumption of residential buildings, in *2010 International Conference on Advances in Energy Engineering* (2010, June), pp. 223–226
13. S. Humeau, T.K. Wijaya, M. Vasirani, K. Aberer, Electricity load forecasting for residential customers: exploiting aggregation and correlation between households, in *2013 Sustainable Internet and ICT for Sustainability (SustainIT)* (2013, Oct), pp. 1–6
14. B. Dong, C. Cao, S.E. Lee, Applying support vector machines to predict building energy consumption in tropical region. Energy Build. **37**(5) 545–553 (2005)
15. Jain, A., Satish, B.: Clustering based short term load forecasting using support vector machines, in *2009 IEEE Bucharest PowerTech* (2009, June), pp. 1–8
16. Mohandes, M.: Support vector machines for short-term electrical load forecasting. Int. J. Energy Res. **26**, 335–345 (2002)
17. A. Rahman, A.D. Smith, Predicting heating demand and sizing a stratified thermal storage tank using deep learning algorithms. Appl. Energy **228**, 108–121 (2018)
18. I. Goodfellow, Y. Bengio, A. Courville, *Deep Learning* (MIT Press, Cambridge, 2016). http://www.deeplearningbook.org
19. T.M. Mitchell, *Machine Learning*, 1st edn. (McGraw-Hill, New York, 1997)
20. F. Farahnakian, T. Pahikkala, P. Liljeberg, J. Plosila, Energy aware consolidation algorithm based on k-nearest neighbor regression for cloud data centers, in *2013 IEEE/ACM 6th International Conference on Utility and Cloud Computing*, pp. 256–259 (2013)
21. J. Han, M. Kamber, *Data Mining: Concepts and Techniques* (Kaufmann, San Francisco [u.a.], 2005)
22. E. Wilson, D.W. Tufts, Multilayer perceptron design algorithm, in *Proceedings of IEEE Workshop on Neural Networks for Signal Processing*, pp. 61–68 (1994, Sept)
23. R.E. Edwards, J. New, L.E. Parker, Predicting future hourly residential electrical consumption: a machine learning case study. Energy Build. **49**, 591–603 (2012)
24. S. Hochreiter, J. Schmidhuber, Long short-term memory. Neural Comput. **9**(8), 1735–1780 (1997)

25. W.A. Jabbar, M.H. Alsibai, N.S.S. Amran, S.K. Mahayadin, Design and implementation of IoT-based automation system for smart home, in *2018 International Symposium on Networks, Computers and Communications (ISNCC)* (2018), pp. 1–6
26. D.P. Kingma, J. Ba, Adam: A method for stochastic optimization (2014). cite arxiv:1412.6980. Comment: Published as a conference paper at the 3rd International Conference for Learning Representations, San Diego, 2015

The Analysis of Different Materials Used for an Electric Car Charger Shell Under the Wind Influence

Catalin Marin Buciuman, Cristian Vilau, Dagmar Cagáňová, and Cristina Miron-Borzan

1 Introduction

Nowadays, the electric vehicles are often used than in the last years due to the advantages that they offer. The increasing of the oil price and also environment pollution determined the needs of using the electric vehicles, especially in cities. The toxic emissions from common car engines lead to global warming and influence the people's health through different diseases. The trend to replace the combustion cars by electric cars is expanded in the transport industry too [1]. Even if a non-renewable energy is used for their charging, the electric vehicles are environment friendly [2]. In the scientific literature, there are many authors who studied the advantages of the electric cars and infrastructure [2–12]. Gnann et al. [3] made a study about potential customers of electric vehicles and their needs. One major question is about the existence and possibility of public charging facilities. This fact is due to the speed of public charging, which is very close to the conventional refueling time. According to the comparison made by Granovskii et al. [4], hybrid and electric cars have many advantages in front of the combustion cars. The source of electricity has a high influence on the economic efficiency and environmental impact.

C. M. Buciuman · C. Vilau · C. Miron-Borzan (✉)
Department of Manufacturing Engineering, Faculty of Machine Building, Technical University of Cluj-Napoca, Cluj, Romania

D. Cagáňová
Institute of Industrial Engineering and Management, Faculty of Materials Science and Technology in Trnava, Slovak University of Technology in Bratislava, Trnava, Slovakia
e-mail: dagmar.caganova@stuba.sk

D. Cagáňová et al. (eds.), *Advances in Industrial Internet of Things, Engineering and Management*, EAI/Springer Innovations in Communication and Computing,
https://doi.org/10.1007/978-3-030-69705-1_9

The interest for electric cars and their infrastructure existed from many years ago. Thus, 25 years ago, Cowan [5] made an analysis of the current state for electric vehicles and the surrounding supporting industries and infrastructures.

The electric distribution network is a large system meant to transfer the electric power from producers to its consumers. The electric power can be produced both from renewable and non-renewable resources [2].

Charging equipment for electric vehicles is classified by the rate at which the batteries are charged. In this moment in the world, there are three types of electric vehicle chargers available on the market: Level 1 chargers (120 V), Level 2 chargers (240 V), Level 3 chargers or DC fast charging (3 phase). The charging principle is the same for all these three charging equipment, using a cable that is connected to a grid. The recharging stations can be installed in parking areas or along the roads [13].

The electric charger has the following main components: the shell, the support of the feeding cable, the feeding cable, and internal components. Usually, the shell is made by metallic materials.

In the last years, it was observed a trend of changing the metal materials with plastic or composite materials. Composite materials, as a combination of more materials, have many advantages in front of the classic materials, also from a mechanical properties point of view (high strength and stiffness, low density). In comparison with metallic alloys, the composite materials maintain for each component material their own properties. Another advantage of composite materials is the lightweight [14].

The finite element analyses (FEA) is a very helpful tool, used in the engineering and designing stage of products. Based on meshing method, the physical system is analyzed. After simulations, the obtained results can influence the decision of the real shape of the products [15, 16].

Taking into consideration different possibilities of materials that can be used in the manufacturing stage of the shell, the purpose of this study was to analyze the behavior of one electric charger shell made from three different materials (aluminum, steel, and composite material), during the wind action, in extreme conditions.

This paper is part of a larger research that is focused on the behavior and properties of different thicknesses for the shell manufactured from different metals and composite materials (glass fiber and epoxy resin, carbon fiber with epoxy resin).

2 Material and Methods

In order to make the simulations, the first objective of this paper was to design a shell for an electric car charger. For obtaining the 3D model of the shell, SolidWorks 2017 software was used. The design of the electric car charger shell is presented in Fig. 1.

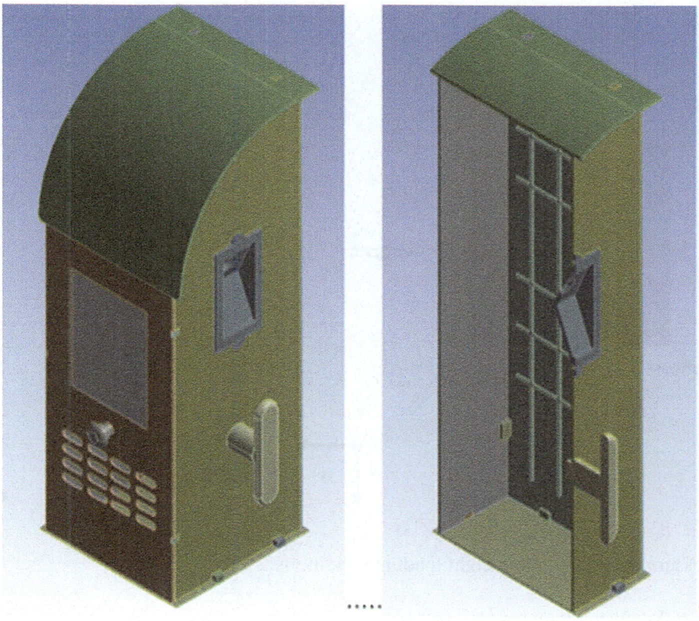

Fig. 1 The proposed design for the electric car charger shell

Table 1 The used materials and their mechanical properties

Material	Density (kg/m^3)	Tensile yield strength (MPa)
Composite material	1451	*X* direction: 513*Y* direction: 513*Z* direction: 50
Aluminum	2770	280
Steel sheet	7850	250

For simulation were considered three types of material: two metallic (steel and aluminum) and one composite material. The wall thickness of the shell was designed at 2 mm for all three materials.

The composite material is made from carbon fiber woven with epoxy resin.

In order to obtain the behavior of the shell during the wind blowing, ANSYS software was used. This software is helping to reduce the costs of physical prototypes, to avoid potential manufacturing, and to analyze the resistance on different loads. Because the wind can blow from all directions, it was considered eight main directions.

For simulation, loading conditions and bearing conditions (fixing with six screws) were required. The mechanical properties of the material were taken from the library of the ANSYS software. Table 1 presents these values.

As was recommended in the scientific literature [17], the pressure of 1000 N/m^2 (representing 100 kg/m^2) was used for the loading conditions. The loadings on the support for the feeding cable and on the backplate (the weight of all internal components) were considered too.

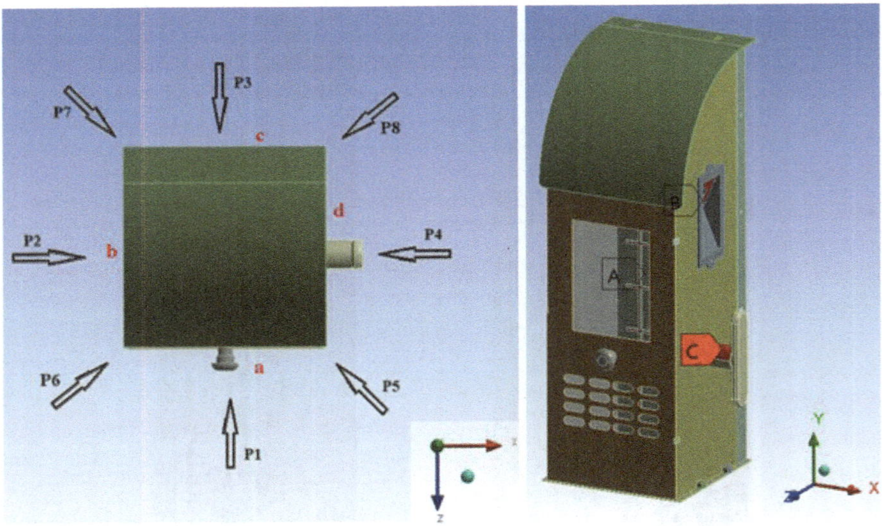

Fig. 2 Pressure directions and weight loading conditions

Table 2 Total deformations (mm)

	Wind direction							
Material	*P*1	*P*2	*P*3	*P*4	*P*5	*P*6	*P*7	*P*8
Aluminium	8.5263	2.5459	3.1503	2.4439	3.0405	2.7331	2.7593	3.1039
Composite material	18.047	6.4022	8.2193	5.983	6.9038	7.6453	7.6881	8.2775
Steel sheet	4.0839	0.93574	1.13114	0.89805	1.1157	1.0082	1.0177	1.1403
	Face							
Aluminium	a	d	c, d	d	a	d	d	a
Composite material	a	d	d	d	a	d	d	d
Steel sheet	a	d	c, d	d	a	d	d	a,d

Figure 2 presents the loading directions for pressure considered for the simulations and loading conditions for the weight of other components (cable and internal components of the charger).

3 Results and Discussions

Following the analysis, maximum values of deformations and stresses were determined with the fixing and bearing conditions previously presented.

Table 2 and Fig. 3 present the total deformations for all three analyzed materials, from all the possibilities of wind direction and also the surface of the shell that was the most affected.

As can be observed in Table 2, the maximum deformation was obtained for the case when the wind is blowing from the *P*1 direction (in the front face of the shell)

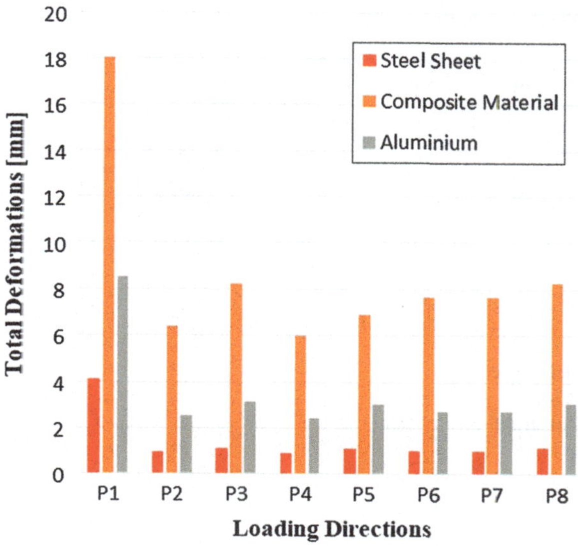

Fig. 3 Comparison between the obtained values for total deformation

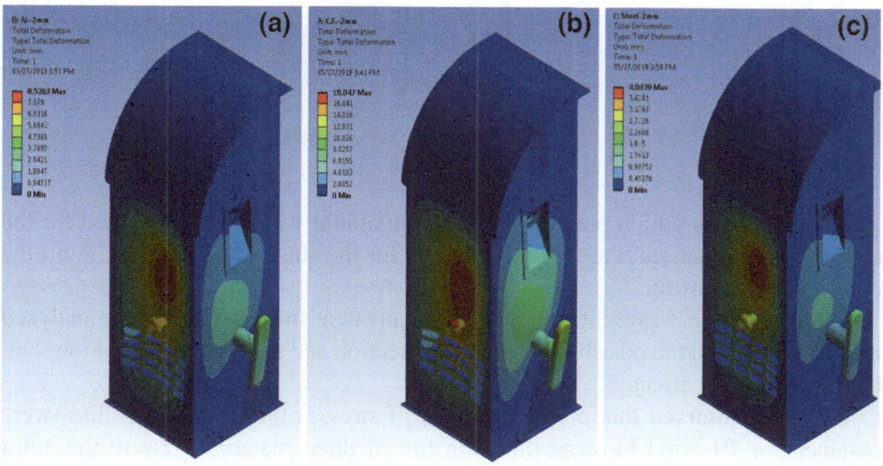

Fig. 4 Maximum total deformations. (**a**) Aluminum; (**b**) composite material; (**c**) steel sheet

and this face was also the most affected for all three materials. These cases can be observed also in Fig. 4 (a—aluminum; b—composite material; c—steel sheet). An explanation for these results can be because in the front face is located the door that have some holes for ventilation. These holes will reduce the surface area and when the wind pressure is acting on a smaller surface, is normal to obtain higher deformations on that surface.

Table 3 Maximum equivalent stress (MPa)

Material	Wind direction							
	*P*1	*P*2	*P*3	*P*4	*P*5	*P*6	*P*7	*P*8
Aluminium	70.303	25.009	50.703	23.685	27.828	39.357	39.352	28.612
Composite material	104.070	34.043	67.885	33.626	41.201	52.893	52.884	43.415
Steel sheet	73.219	24.922	51.144	23.86	28.642	39.697	39.692	29.355
	Face							
Aluminium	a	d	c	d	a, d	d	d	d,a
Composite material	a	d	c	d	a, d	d	d	d,a
Steel sheet	a	d	c	d	a, d	d	d	d,a

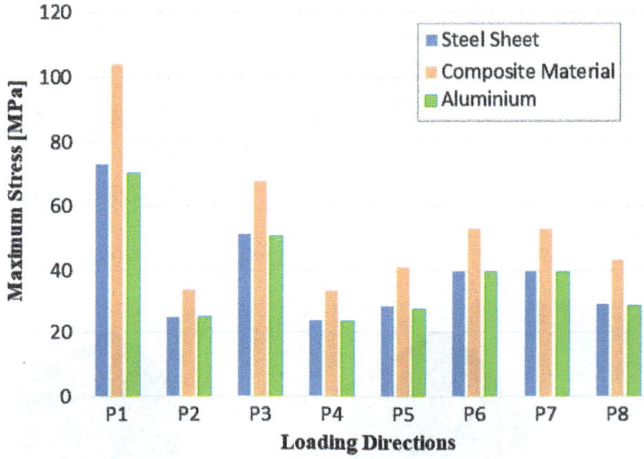

Fig. 5 Comparison between the obtained values for maximum equivalent stress

From the deformation point of view the maximum values were obtained for the composite material and the minimum values for the sheet steel material, in all the cases of wind blowing.

Table 3 and Fig. 5 present the maximum equivalent stresses for all three analyzed materials, from all the possibilities of wind direction and also the surface of the shell that was the most affected.

It can be observed that also in the case of stresses the maximum values were obtained for *P*1 wind blowing direction for all three materials, due to the holes from the front door. The maximum values resulted for the carbon fiber composite material, followed by the sheet steel material.

Figure 6 presents the maximum equivalent stress for the analyzed materials (a—aluminum; b—composite material; c—steel sheet).

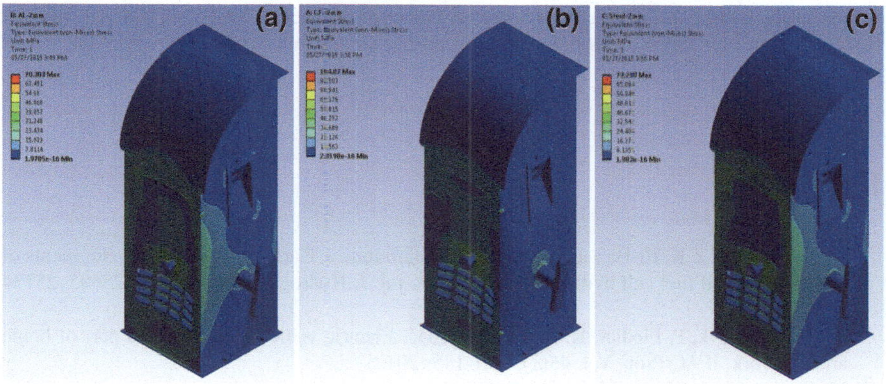

Fig. 6 Maximum equivalent stresses. (**a**) Aluminum; (**b**) Composite material; (**c**) Steel Sheet

4 Conclusions

The electric vehicles started to be often used than in the last years due to the advantages that they offer. The growing of electric vehicles number involves also the developing of the charging infrastructure.

Nowadays, the electric charger shell is made from a metallic material. The paper analyzed three different materials (steel, aluminum, and a composite material from carbon fiber woven and epoxy resin) that can be used in the manufacture of an electric car charger shell, in terms of the stresses and deformations occurring under wind action (in extreme conditions). For the analyses, ANSYS software was used and deformations and equivalent stresses for all the proposed materials were obtained.

The results showed that for all the materials the maximum deformations were obtained in the case when the wind was blowing from the face of the shell and the front door was the most affected in this situation. For the composite material was obtained the maximum value for deformation 18,047 mm, a higher value that it was obtained for the metallic materials.

From equivalent stresses point of view, the maximum values were obtained for the same case, when the wind is blowing from the front direction ($P1$). For this case, the maxim obtained value was for the composite material, 104 MPa. The difference between the two analyzed metallic materials was very small (approx. 3 MPa).

Comparing the values for the equivalent stresses it could be observed that all the values for the two metallic materials were very close.

Following the analyses, it can be concluded that all three materials have good behavior in the case of extreme conditions and will not fail under the loadings.

The study showed that all three materials are adequate to be used in the manufacture of the electric car charger shell from the mechanical behavior point of view.

Acknowledgement This chapter was prepared as part of the KEGA Nr. 030STU-4/2018 project E-platform for improving collaboration among universities and industrial enterprises in the area of education.

References

1. T. Wilberforce, Z.K. El-Hassan, A. Makky, A. Baroutaji, J. Carton, A. Olabi, Developments of electric cars and fuel cell hydrogen electric cars. Int. J. Hydrog. Energy **42**(40), 25695–25734 (2017)
2. V. Kaczmarczyk, P. Fiedler, R. Štohl, Z. Bradáč, Electric vehicles charger as a part of home area network. IFAC Proc. Vol. **45**(7), 180–185 (2012)
3. M. Granovskii, I. Dincer, M. Rosen, Economic and environmental comparison of conventional, hybrid, electric and hydrogen fuel cell vehicles. J. Power Sources **159**(2), 1186–1193 (2006)
4. T. Gnanna, S. Funkea, N. Jakobssonb, P. Plötza, F. Spreib, A. Bennehagc, Fast charging infrastructure for electric vehicles: Today's situation and future needs. Transp. Res. D **62**, 314–329 (2018)
5. R. Cowan, S. Hultén, Escaping lock-in: The case of the electric vehicle. Technol. Forecast. Soc. Chang. **53**(1), 61–79 (1996)
6. T. Wilberforce, Z. El-Hassan, F.N. Khati, A. Al Makky, A. Baroutaji, J. Carton, A. Olabi, Developments of electric cars and fuel cell hydrogen electric cars. Int. J. Hydrog. Energy **42**(40), 25695–25734 (2017)
7. J. Fontaínhas, J. Cunha, P. Ferreira, Is investing in an electric car worthwhile from a consumers' perspective? Energy **115**, 1459–1477 (2016)
8. J. Ferreira, V. Monteiro, J. Afonso, A. Silva, Smart electric vehicle charging system, in *EEE Intelligent Vehicles Symposium (IV)*, Baden-Baden, Germany (2011)
9. E. Tate, M. Harpster, P. Savagian, The electrification of the automobile: From conventional hybrid, to plug-in hybrids, to extended-range electric vehicles, in *SAE International World Congress* (2008)
10. Y.A. Alhazmi, H.A. Mostafa, M.M.A. Salama, Optimal allocation for electric vehicle charging stations using Trip Success Ratio. Int. J. Electr. Power Energy Syst. **91**, 101–116 (2017)
11. T. Gnann, D. Goldbach, N. Jakobsson, P. Plötz, A. Bennehag, F. Sprei, A model for public fast charging infrastructure needs, in *EVS29 International Battery, Hybrid and Fuel Cell Electric Vehicle Symposium*, Montreal, Canada (2016)
12. P. Morrissey, P. Weldon, M. O'Mahony, Future standard and fast charging infrastructure planning: an analysis of electric vehicle charging behaviour. Energy Policy **89**, 257–270 (2016)
13. G. Buja, M. Bertoluzzo, N. Mude, Design and experimentation of WPT charger for electric city car. IEEE Trans. Ind. Electron. **62**(12), 7436–7446 (2015)
14. F.C. Campbell, *Introduction to Composite Materials. Structural Composite Materials* (ASM International, Materials Park, OH, 2010)
15. D.V. Hutton, *Fundamentals of Finite Element Analysis* (Elizabeth A. Jones Publishers, New York, 2004)
16. E. Sabau, A. Popescu, C. Vilau, Mechanical behavior of composite materials using the finite element analysis. MATEC Web Conf. **137**(08006), 1–7 (2017)
17. L. Negi, *Design of Steel Structures* (Tata McGraw-Hill Education, New Delhi, 1997). ISBN: 10-0-07-462305-2

Improvement of Hydraulic Characteristics for Impellers Using the Finite Volume Analysis

Maryna Demianenko, Oleksandr Starynskyi, Bohdan Vashyst, Ivan Pavlenko, Oleksandr Liaposhchenko, and Vitalii Ivanov

1 Introduction

One of the most crucial pump cards is the head–capacity curve, which is the dependence between the pump head and the volumetric flow. The shape of this card depends on the pump type, its design features, and other parameters. There are two types of the head–capacity curve: theoretical and experimental. The theoretical one is determined by applying the fundamental equations of hydrodynamics with the corresponding assumptions and simplifications; the experimental one is based on experimental data. Both have the following disadvantages: for the first—it accurately reflects only the qualitative relationship between the main parameters; for the second—it is resource-intensive. One of the main problems of the pump operation process is cavitation (gas bubbles formation process due to a sharp pressure change caused by the fast increase of the fluid velocity). The above process is characterized by the cavitation number, defined as:

$$\mathrm{Ca} = \frac{p - p_{\mathrm{v}}}{\frac{1}{2}\rho U^2}, \tag{1}$$

where p—flow pressure, p_{v}—vapor pressure, $\frac{1}{2}\rho U^2$—dynamic pressure.

The primary way for cavitation avoidance is to minimize the underpressure value in the critical areas, which can partially be achieved by increasing the ambient pressure. In particular, the main thing for the cavitation preventing is to ensure

M. Demianenko · O. Starynskyi · B. Vashyst · I. Pavlenko · O. Liaposhchenko · V. Ivanov (✉)
Sumy State University, Sumy, Ukraine
e-mail: ivanov@tmvi.sumdu.edu.ua

D. Cagáňová et al. (eds.), *Advances in Industrial Internet of Things, Engineering and Management*, EAI/Springer Innovations in Communication and Computing,
https://doi.org/10.1007/978-3-030-69705-1_10

suction pressure, which provides the overcoming fluid flow without hydraulic losses in the suction line and the suction chamber. The inducer can be used for the above purpose.

Typically, cavitation is considered separately from the phase–temperature transition due to its processing speed, which leads to the assumption of thermal equilibrium unacceptably. The mass transfer is determined by a purely mechanical interaction (the pressure difference at the liquid–vapor interface) and not dependent on thermal interaction in the simplest cavitation model. Currently, many types of research are aimed at models that consider both effects.

2 Literature Review

The reliability of hydraulic motors is a topical engineering problem. A specific design method was proposed to increase reliability based on the developed mathematical tools through simulating changes in a technical state of the rotors in a hydraulic motor [1]. The study [2] introduces the design schemes, a mathematical tool, and a calculation algorithm that were developed to justify the angular arrangement of the moving distributor. In the research study [3], a design model, a mathematical apparatus, and a calculation algorithm were developed to solve the problem of manufacturing workability improvement. The reliability of pneumatic and hydraulic machines is very acute. Experimental studies on a specially designed experimental setup with a supercharger model proved that the ejection ratio for the solid particle is twice that of the jet ejectors [4]. The experimental studies proved the possibility of improvement of cavitation erosion characteristics in the centrifugal inducer stage with the inducer bush [5]. Due to the rapid development of modern computer technology, nowadays, many computer programs based on the finite volume method are used for simulation hydrodynamics in multiple channels. Modeling viscous flow hydrodynamics using state-of-the-art software products is widely used by leading foreign researchers to solve pump engineering problems [6–13]. Engineering calculations are complicated and time-consuming tasks, so they should be automated and integrated with computer-aided systems. It can ensure the implementation of the Industry 4.0 strategy [14–16]. The right and the correct choice of the turbulence and cavitation models was made by the consideration of a number of the scientific works of this direction. One of the most widely used software is ANSYS CFX [17–19], which allows you to study the fluid flow in the pump and obtain the head–capacity curve. It should be noted that the head–capacity curve obtained from numerical simulation has a good correlation with experimental results subject to a provided computational model. So, it can be used for the optimization of the introduced geometry before pump production. The parametric method for constructing geometry is necessary. The Design Modeler unit, in particular, Blade Modeler [20], is used for profiling blades. ANSYS Blade Modeler is a powerful specialized tool for working with the geometry of machine blades; it allows you to create parametric models of the blade ring using both the available theoretical drawings and the CAD model. The integrated design modules allow creating the geometry of a centrifugal machine from scratch with the subsequent construction of a parametric three-dimensional geometry model for analysis and optimization.

The calculation of the fluid flow is carried out by numerically solving a system of equations describing the most general case of fluid motion—the Navier–Stokes equations and the continuity equation. The Reynolds equations are used with the closure by the turbulence models in ANSYS CFX—the widely used SST turbulence model. According to this model, the system of equations of fluid motion is supplemented by two differential equations describing the transfer, respectively, of the kinetic energy of turbulence k and the dissipation rate ε (for the main flow) and ω (for boundary layer).

$$\frac{\partial}{\partial t}(\rho k)+\frac{\partial}{\partial x_j}\left(\rho \overline{u_j} k\right)=\frac{\partial}{\partial x_j}\left(\Gamma_k \frac{\partial k}{\partial x_j}\right)+P_k-\rho\varepsilon, \tag{2}$$

$$\frac{\partial}{\partial t}(\rho\varepsilon)+\frac{\partial}{\partial x_j}\left(\rho \overline{u_j}\varepsilon\right)=\frac{\partial}{\partial x_j}\left(\Gamma_\varepsilon \frac{\partial \varepsilon}{\partial x_j}\right)+\frac{\varepsilon}{k}\left(C_{\varepsilon 1} P_k-\rho C_{\varepsilon 2}\varepsilon\right), \tag{3}$$

where $P_k = -\rho\overline{u_i' u_j'}\frac{\partial \overline{u_i}}{\partial x_j}$—energy expression term k.

The cavitation process in the flow part of the centrifugal stages is considering using a combination of the Reilly–Plessis model implemented to the multiphase flow and interphase mass transfer model. A uniform multiphase model is usually used for flows with cavitation.

The Reilly–Plessis equation is the basis for the equation determining the rate of formation of steam and condensate; it describes the growth of gas bulbs in a liquid, defined as:

$$R_{\mathrm{B}}\frac{a^2 R_{\mathrm{B}}}{dt^2}+\frac{3}{2}\left(\frac{dR_{\mathrm{B}}}{dt}\right)^2+\frac{2\sigma}{\rho_{\mathrm{f}} R_{\mathrm{B}}}=\frac{p_{\mathrm{v}}-p}{\rho_{\mathrm{f}}}, \tag{4}$$

where R_{B}—bubble radius, p_{v}—bubble pressure, p—fluid pressure around the bubble, ρ_{f}—fluid density, σ—surface tension coefficient between liquid and vapor.

Given the above, the main aim of this paper is to construct the parametric geometry of the elements centrifugal pump and determine the main operation parameters by using the finite volume method.

3 Research Methodology

The main elements creation of the pump centrifugal stage, namely the impeller, the booster, the inducer, and the inlet distribution part, was carried out using the Blade Modeler software package (Fig. 1).

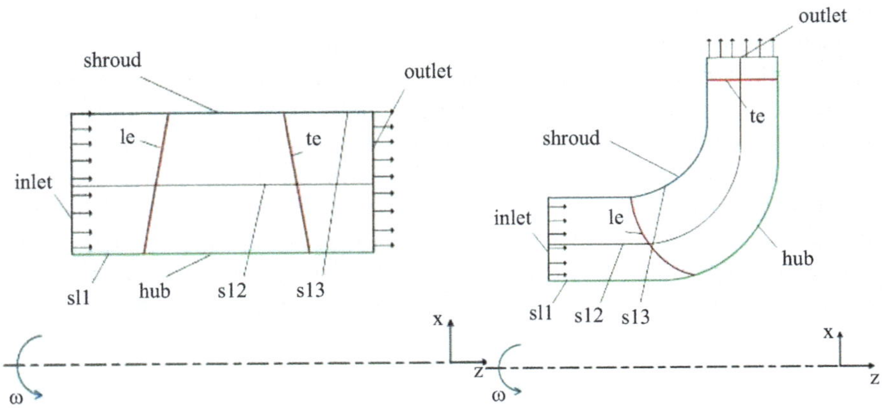

Fig. 1 The geometry designing elements in the Blade Modeler software

Table 1 The geometry designing elements in the Blade Modeler software

Inlet	Inlet part
Outlet	Outlet part
Hub	Internal streamline
Shroud	Outer streamline
le	Entrance edge of the blade
te	Exit edge of the blade
sl	Streamline

The model designing of the interlobe space using the Blade Modeler has certain specifics. The parametric geometry designing using the Blade Modeler is implemented using the seven elements shown in Table 1.

The interlobe space model of almost any blades configuration may be generated using these seven elements. It should be noted that the above elements are designing in a meridional projection.

By default, the impeller's rotation is carried out only *OZ*-axis around, the positive direction of the working fluid flow coincides with the *OZ*-axis direction, the positive direction of rotation of the flowing part is clockwise, viewed from the working area entrance side.

At the first stage of the geometry designing the primary elements sketches (inlet, outlet, hub, shroud) was created. The dimension lines and their sizes were assigned for each of the elements that were specified, which allows changing them if it is necessary. Figure 2 shows the resulting interlobe space contours.

The second stage was the interlobe space generation and blades generation using the FlowPath and Blade tool, respectively. In addition to the meridional projection configuration, the blades are also characterized by angles, namely the wrap angles (φ) and blade installation angles at the inlet and outlet edges (βb). The Blade Modeler provides blades profiling in two modes: the wrap angles and the angle

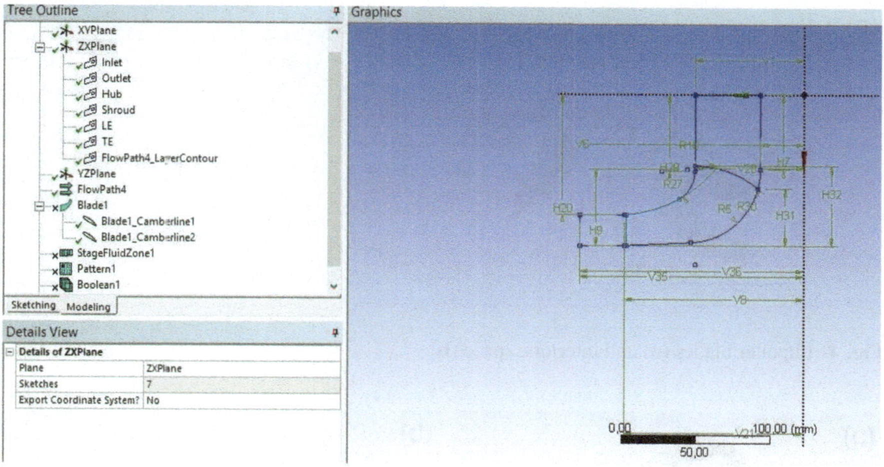

Fig. 2 The interlobe space contours

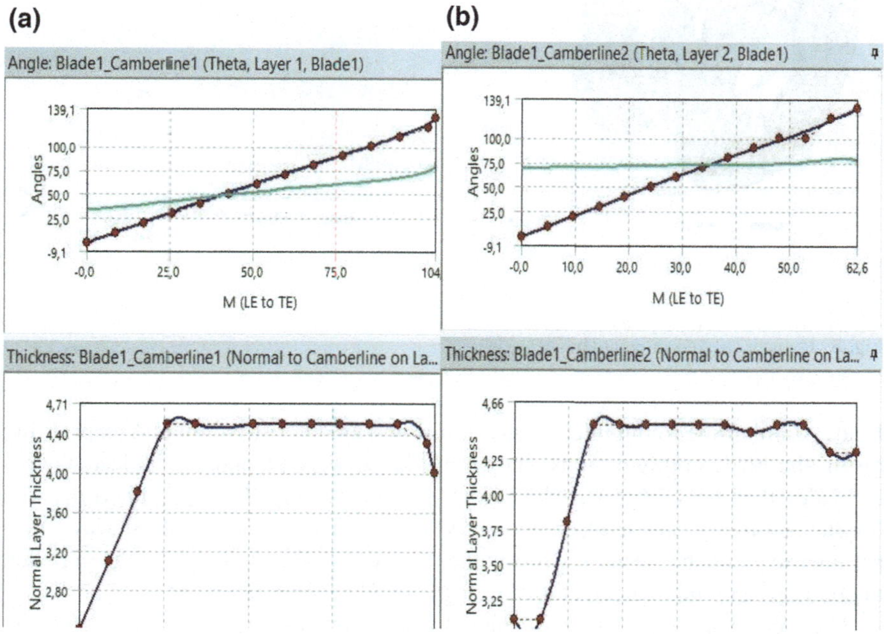

Fig. 3 Blade_Camberline: (**a**) internal streamline; (**b**) outer streamline

of installation, and these parameters are interconnected. The value changing of one parameter in a certain way changes the second parameter value. Blade profiling was performed using Blade_Camberline. It should be noted that for profiling the blades along the internal streamline (hub) along the streamline (SL) and the outer streamline (shroud), there are individual Blade_Camberline. Figure 3 shows

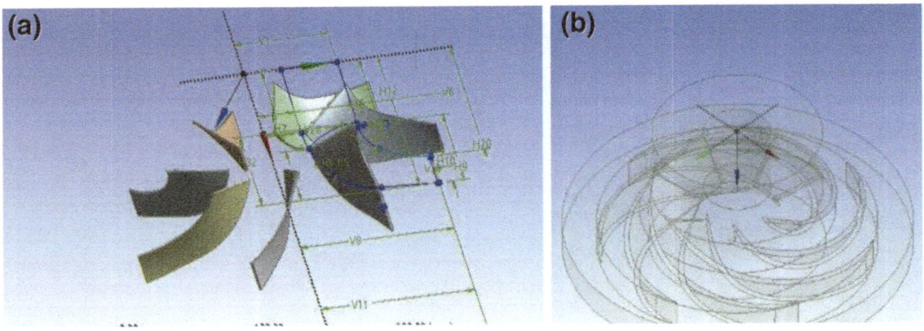

Fig. 4 Impeller blades (**a**) and interlobe space (**b**)

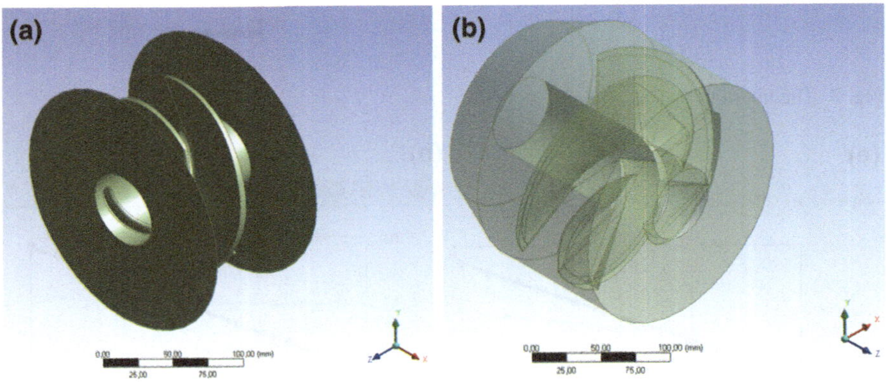

Fig. 5 Booster (**a**) and inducer (**b**)

Blade_Camberline for an internal streamline and an outer streamline. As a profiling result, the blade configurations, shown in Fig. 4, were obtained. The last stage is to blades thickness imparts, which was also done using Blade_Camberline. The booster geometry and inducer geometry were generated in the same way. As a parametric geometry designing result, the three-dimensional models of the impeller, the booster, and the inducer, shown in Figs. 4 and 5, were generated.

The rectilinear region of the 3D model with a flow distributor is shown in Fig. 6.

After creating 3D models, the mesh for all parts of the centrifugal stage was developed using the ANSYS Meshing mesh generator (Fig. 7). The booster mesh and the inducer mesh are presented in Figs. 8 and 9, respectively.

The mesh was generated for each flow part separately, after which they were combined in the CFX pre-processor into one in the following sequence: the rectilinear region, then the booster, the inducer, and the impeller (Fig. 10). In the single-phase flow simulation with non-cavitation as the boundary conditions at the

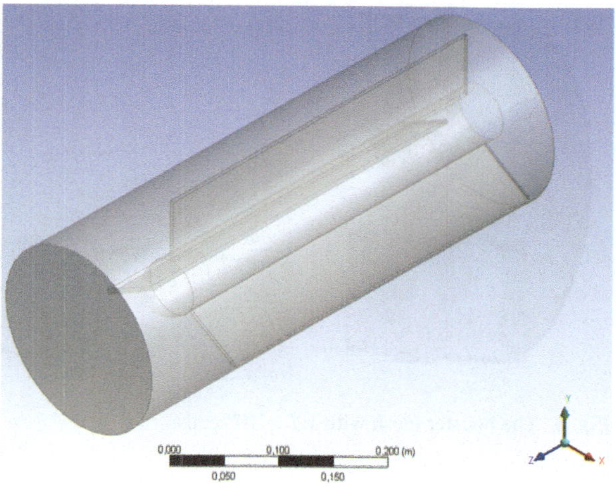

Fig. 6 The rectilinear region of the 3D model with a flow distributor

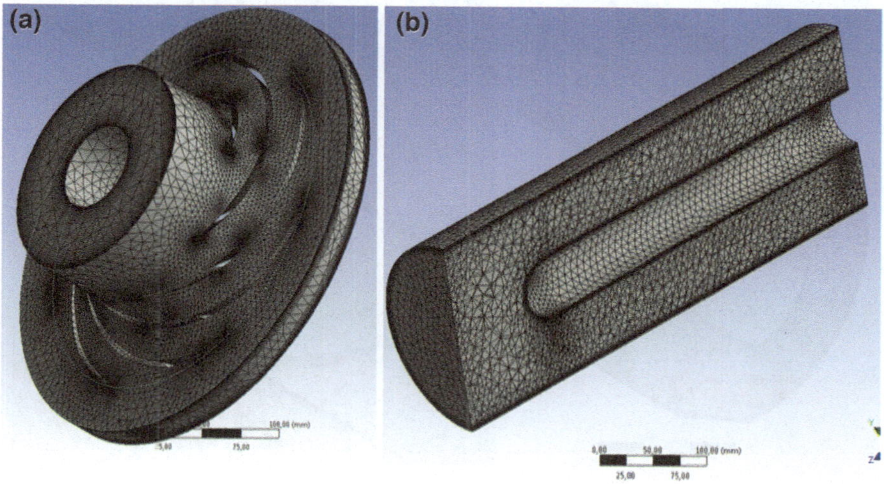

Fig. 7 The impeller mesh with 2.9×10^6 cells (**a**) and the rectilinear region mesh (**b**)

computational domain inlet was set mass flow rate, at the computational domain outlet—static pressure equals 1 MPa. The boundary condition type was configured as "opening" since it was assumed that backflows at the computational domain exit existed.

For all computational domain walls, the velocity which is equal to zero (the "sticking" condition) was specified. The walls were roughness; the mean arithmetic deviation of the profile (R_a) was 6.3 μm. The interface areas of the interaction verges of the rotor and stator elements were determined. The interface type was specified as a "frozen rotor", which assumed parameter averaging at time. The input parameters are shown in Table 2.

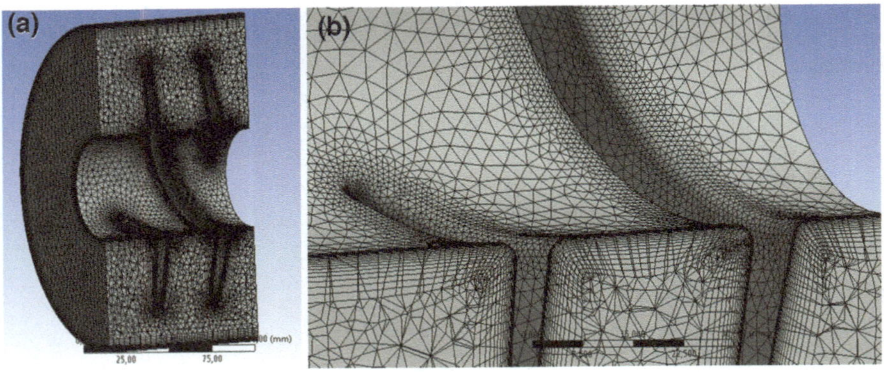

Fig. 8 The booster mesh with 1.7×10^6 cells

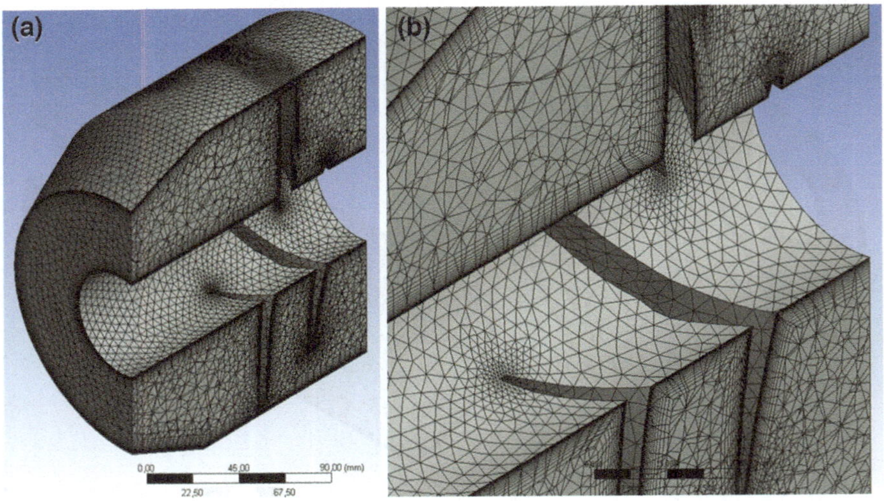

Fig. 9 The inducer mesh with 1.6×10^6 cells

4 Results

The values of velocity and pressure in the calculated areas were obtained as a result of a numerical simulation. Figures 11 and 12 show the pressure fields on the surfaces of a centrifugal impeller, a booster, and an inducer.

The presence of zones with reduced pressure can be noted after consideration of the total pressure contours (Figs. 11 and 12). These are the zones of the most probable occurrence of cavitation cavities. Figures 13 and 14 show velocity vectors in the entire computational domain and separately around the booster and inducer.

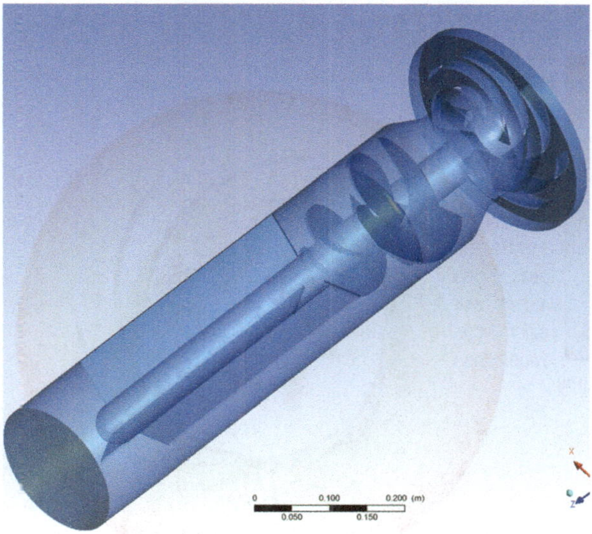

Fig. 10 The centrifugal stage flow part

Table 2 Input parameter for the flow hydrodynamics modeling in the centrifugal stage

Rotation frequency, n	2940 rpm
Density, ρ	997 ks/m^3
Saturated pressure at $t = 25$ °C	3196 pa
Barometric pressure	101,325 pa

The main aim of the flow study in a centrifugal pump is the analysis of the flow structure at the outlet from the inducer. The dimensionless values of the flow parameters were used for the analysis. Meridian component of absolute velocity:

$$\overline{V}_{\mathrm{m}} = \frac{V_{\mathrm{m}}}{\left(\overline{V}_{\mathrm{mcp}}\right)_G}. \tag{5}$$

Circumferential component of velocity:

$$\overline{\upsilon}_u = \frac{\upsilon_u}{u_{\mathrm{pc}}}. \tag{6}$$

where u_{pc}—the circumferential velocity at the outer diameter.

Figures 15, 16 and 17 show the velocity fields distribution at the outlets of the booster and inducer and pressure field.

The pump head and flow, as well as the power and efficiency, were determined as a result of the calculation. The pressure of the centrifugal stage was calculated as the ratio of the pressure difference at the outlet and the inlet of the flow density to

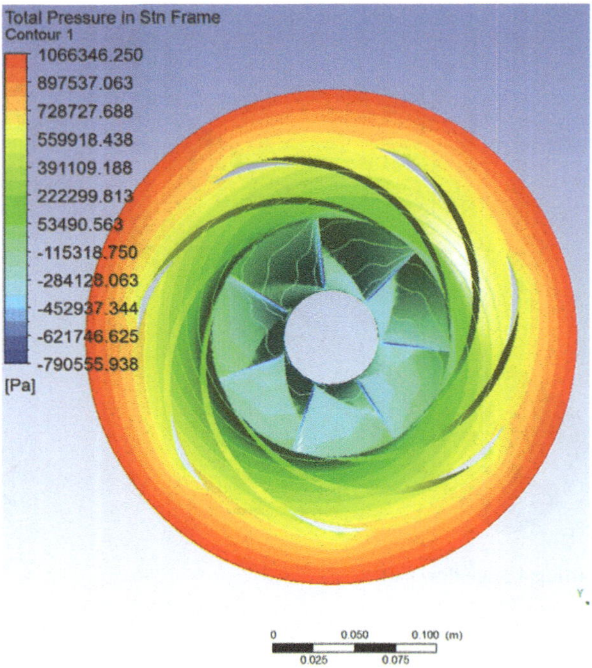

Fig. 11 The distribution of the total pressure on the surface of the centrifugal impeller

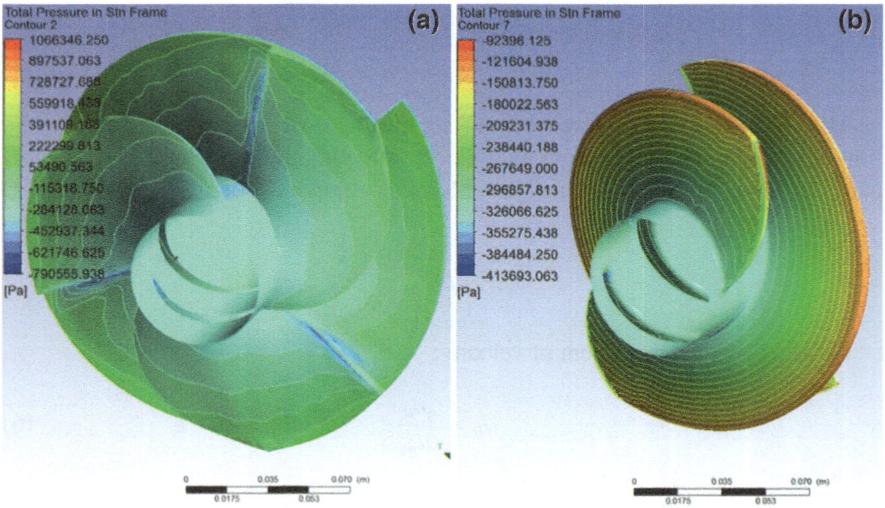

Fig. 12 The distribution of the total pressure on the surface of the inducer (**a**) and the booster's surface (**b**)

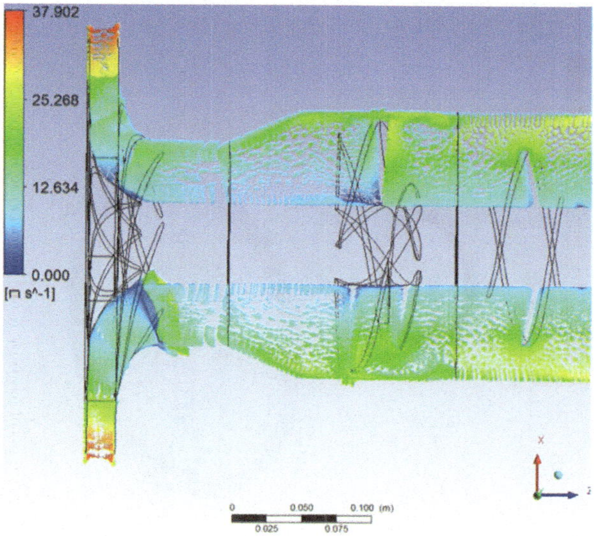

Fig. 13 Velocity vectors in the computational domain

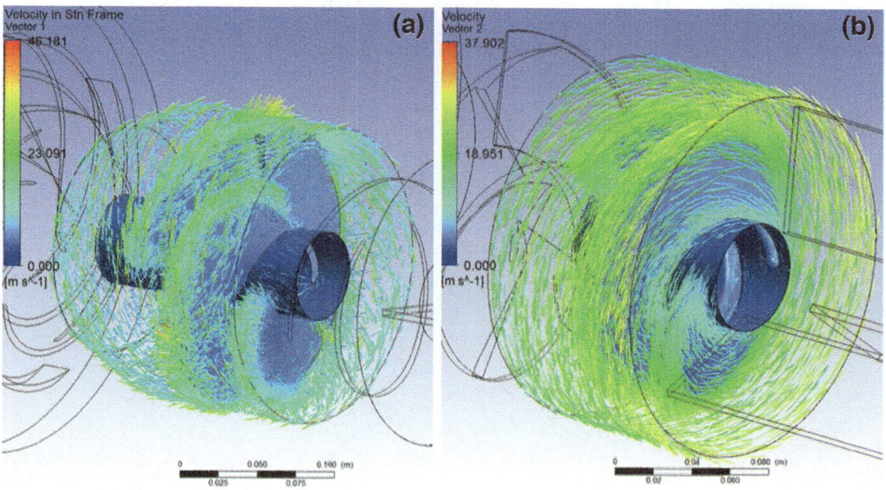

Fig. 14 Velocity vectors around the inducer (**a**) and the booster (**b**)

the acceleration of gravity and amounted to $H = 43.6$ m. The feed of the centrifugal stage is equal to $Q = 236.5$ m^3/s. The pump power was determined by the torque and amounted to $N = 2.86 \times 10^4$ W at a frequency of 2940 rpm. By the following dependence, the pump efficiency was calculated ($\eta = 0.98$):

$$\eta = \frac{\rho g Q H}{3.6 \times 10^3 \cdot N}. \tag{7}$$

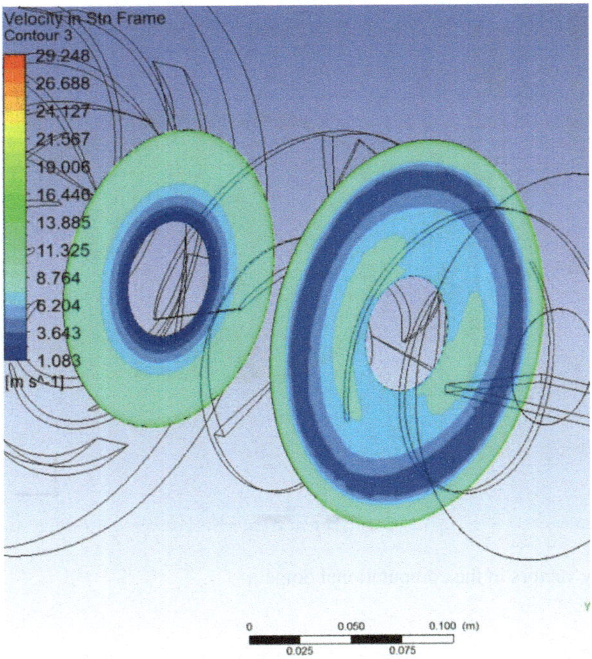

Fig. 15 Velocity distribution at the outlet of the inducer and booster

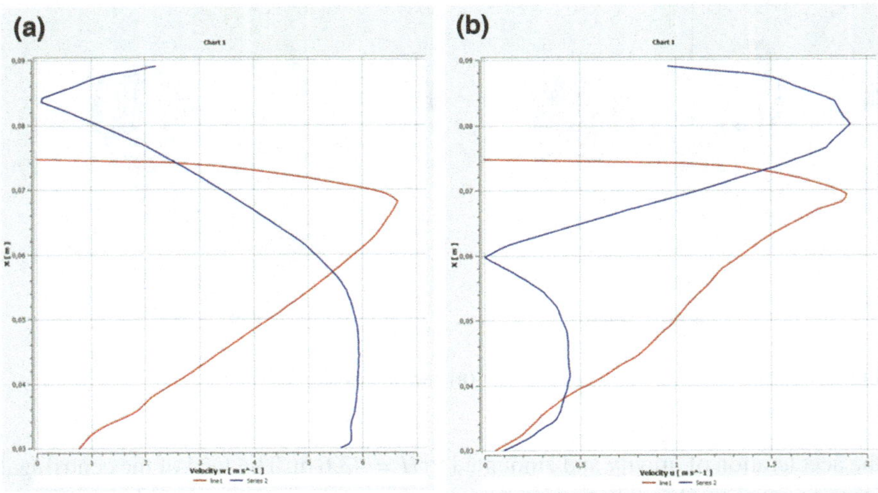

Fig. 16 Distribution graph of the meridian (w) velocity component (**a**) and circumferential (u) component of velocity (**b**) at the outlet of the inducer

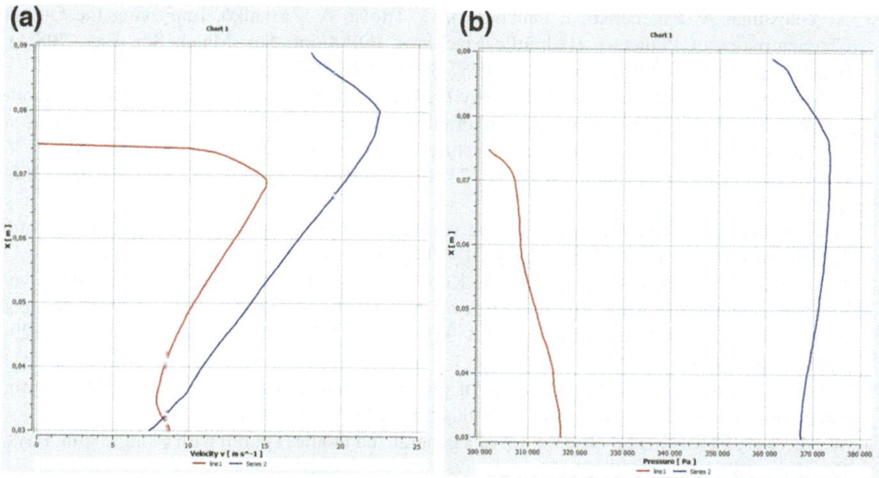

Fig. 17 Distribution graph of the radial (v) velocity component (**a**) and pressure (P) at the outlet of the booster (**b**)

5 Conclusions

The parametrized geometry of the centrifugal stage main elements, namely the impeller, the booster, and the inducer, was done using the Blade Modeler module of the ANSYS software package, for the blades profiling, and given them thickness the Blade_Camberline tool was used. A flow hydrodynamics simulation in the centrifugal stage using the finite volume method was carried out. Based on the simulation results, the individual element's local characteristics of the centrifugal stage were determined, namely the pressure difference on the blade's surface, the flow speed directions of the inducer interlobe space, and the booster inducer. The meridional, axial, and radial velocity components of the inducer are also determined. The centrifugal stage main characteristics, namely the head ($H = 43.6$ m) and the flow rate ($Q = 236.5$ m^3/s), the power ($N = 2.86 \times 10^4$ W) and efficiency ($\eta = 0.98$), was determined. It was noted that the simultaneous use of an inducer and booster reduces pressure drops and, as a result, cavitation.

Acknowledgments This research is partially funded by Center for Industrial Engineering (Sumy State University), International Association for Technological Development and Innovations, and research grants No. 0117U003931 and No. 0120U102036 (Ministry of Education and Science of Ukraine).

References

1. A. Panchenko, A. Voloshina, I. Panchenko, O. Titova, A. Pastushenko, Reliability design of rotors for orbital hydraulic motors. IOP Conf. Ser. Mater. Sci. Eng. **708**(1), 012017 (2019). https://doi.org/10.1088/1757-899X/708/1/012017

2. A. Voloshina, A. Panchenko, I. Panchenko, O. Titova, A. Zasiadko, Improving the Output Characteristics of Planetary Hydraulic Machines. IOP Conf. Ser. Mater. Sci. Eng. **708**(1), 012038 (2019). https://doi.org/10.1088/1757-899X/708/1/012038
3. A. Voloshina, A. Panchenko, O. Boltyansky, O. Titova, Improvement of manufacture workability for distribution systems of planetary hydraulic machines, in *Advances in Design, Simulation and Manufacturing II. DSMIE 2019*, Lecture Notes in Mechanical Engineering, ed. by V. Ivanov et al., (Springer, Cham, 2020), pp. 732–741. https://doi.org/10.1007/978-3-030-22365-6_73
4. A. Rogovyi, Energy performances of the vortex chamber supercharger. Energy **163**, 52–60 (2018). https://doi.org/10.1016/j.energy.2018.08.075
5. P. Tkach, A. Yashchenko, O. Gusak, S. Khovanskyy, V. Panchenko, I. Grechka, Improvement of cavitation erosion characteristics of the centrifugal inducer stage with the inducer bush. Eastern-Eur. J. Enterp. Technol. **4**(8–94), 24–31 (2018). https://doi.org/10.15587/1729-4061.2018.139392
6. F. Shi, H. Tsukamoto, Numerical study of pressure fluctuations caused by impeller-diffuser interaction in a diffuser pump stage. J. Fluids Eng. **123**(3), 466–474 (2001)
7. W. Ma et al., Experimental observation of a generalized thouless pump with a single spin. Phys. Rev. Lett. **120**(12), 120501 (2018)
8. H. Wang, H. Tsukamoto, Fundamental analysis on rotor-Stator interaction in a diffuser pump by vortex method. J. Fluids Eng. **123**(4), 737–747 (2001)
9. J. Sun, H. Tsukamoto, Off-design performance prediction for diffuser pumps. Proc. Inst. Mech. Eng. A J. Power Energy **215**(2), 191–201 (2001)
10. H. Tsukamoto, H. Ohashi, Transient characteristics of a centrifugal pump during starting period. J. Fluids Eng. **104**(1), 6–13 (1982)
11. W. Qin, H. Tsukamoto, Theoretical study of pressure fluctuations downstream of a diffuser pump impeller—part 1: Fundamental analysis on rotor-Stator interaction. J. Fluids Eng. **119**(3), 647–652 (1997)
12. S.-M. Kwon, H.S. Kang, J.-H. Shin, Rotor profile design in a hypogerotor pump. J. Mech. Sci. Technol. **23**(12), 3459–3470 (2009)
13. H.J. Kong, S.J. Kang, K.S. Yun, H.J. Lim, An experimental study on the cooling and heating performance of a residential ground source heat pump system. Korean J. Air-Cond. Refrig. Eng. **25**(3), 156–163 (2013)
14. S. Saniuk, A. Saniuk, D. Cagáňová, Cyber industry networks as an environment of the Industry 4.0 implementation. Wirel. Netw. (2019). https://doi.org/10.1007/s11276-019-02079-3
15. N. Horňáková, L. Jurík, H. Hrablik Chovanová, D. Cagáňová, D. Babčanová, AHP method application in selection of appropriate material handling equipment in selected industrial enterprise. Wirel. Netw. (2019). https://doi.org/10.1007/s11276-019-02050-2
16. V. Švač, D. Cagáňová, Managerial skills for innovation support. Mobile. Netw. Appl. (2020). https://doi.org/10.1007/s11036-020-01517-3
17. P. Dupont, E. Casartelli, Numerical prediction of the cavitation in pumps, in *Symposia and General Papers, Parts A and B*, vol. 2 (2002)
18. R. Ferman et al., Boiler feed pump rerate for increased head and reduced cavitation, in *3rd ASME Pumping Machinery Symposium*, Vancouver, Canada (1997)
19. J.P. Franc, The Rayleigh-Plesset equation: A simple and powerful tool to understand various aspects of cavitation, in *Fluid Dynamics of Cavitation and Cavitating Turbopumps*, CISM International Centre for Mechanical Sciences, ed. by L. d'Agostino, M. V. Salvetti, vol. 496, (Springer, Vienna, 2007)
20. L. Shabliy, A. Cherniaev, Optimization of gas turbine compressor blade parameters for gas-dynamic efficiency under strength constraints, in *Proceedings of the 4th International Conference on Simulation and Modeling Methodologies, Technologies and Applications* (2014)

Management Decision-Making for Logistics Systems Using a Fuzzy-Neural Simulation

Dmitriy Muzylyov, Natalya Shramenko, and Vitalii Ivanov

1 Introduction

It is necessary to have data on future transportation volumes in order for management decisions to be correctly developed during the planning phase. At which the forecast should have an error of not more than 5%. Information on grain transshipment volumes through the offshore terminal is key for organizing the transportation process based on research object essence.

An increase in cargo turnover of Ukrainian seaports is evidence of growth dynamics in the future period compared to previous years according to their operation results during 2019 [1]. At the same time, certain loads may arise on logistics chain main links, especially in hubs of interaction between different kinds of transport. Therefore, there is a need to ensure the continuity and efficiency of transport processes.

Agricultural goods share transferred through Ukrainian seaports amounted to about 37.4% of the total volume of transshipment (160 million tons in 2019). This information was obtained according to statistical information, during the last reporting year. These were primarily cereals exported to European Union countries and other world regions.

The trend towards an increase in cargo flows processed by Ukrainian seaports is observed based on recent statistical data. Therefore, this aspect indicates an increase in storage capacity and processing capacity by overloading mechanisms [2]. However, significant investments and long payback periods are required for

D. Muzylyov (✉) · N. Shramenko
Kharkiv Petro Vasylenko National Technical University of Agriculture, Kharkiv, Ukraine

V. Ivanov
Sumy State University, Sumy, Ukraine
e-mail: ivanov@tmvi.sumdu.edu.ua

D. Cagáňová et al. (eds.), *Advances in Industrial Internet of Things, Engineering and Management*, EAI/Springer Innovations in Communication and Computing,
https://doi.org/10.1007/978-3-030-69705-1_11

Fig. 1 Overloading of grain cargoes from trucks to ship using mobile loading mechanism

increasing these indicators of transport hub operations. Therefore, the search for alternative ways of interacting between different kinds of transport is more pragmatic for logistics operators during agricultural cargo transshipments [3].

As a result, the direct option of uploading a ship from trucks is increasingly used in European ports (Fig. 1).

The financial and time costs of the company will be minimal, using a direct interaction option. Forwarders should plan the process of interaction between road and sea kinds of transport considering the exact forecast to avoid port malfunctions during grain transshipment. Error-free forecasting of transshipment volumes allows optimizing works of all participants in grain supply chains [4]. However, the random nature of transportation processes complicates the forecasting and often leads to unacceptable error.

It is recommended to use integrated solutions to organize and plan production processes based on the latest researches in Industry 4.0. Such a concept involves the significant use of smart technologies, which helped to simplify distribution processes while sharing production resources. Mentioned-above significantly increases the efficiency of management decision-making, especially in conditions of difficulty in predicting future volumes [5, 6].

The mathematical tool of self-learning and self-tuning neural networks has become more often used recently as an adequate solution to indicated problems [7]. This approach allows for obtaining an adequate forecast of cargo transshipment volumes through port terminals. Application of approaches based on evolutionary principles allows us to significantly save time when making the right management decisions in conditions of constant distribution of material flow [8, 9] or multi-criteria [10] based on conceptual solutions according to the Industry 4.0 principles. The presented variation of simulation of various parameters of material flow allows achieving optimal solutions when researchers have a small amount of information. It is also suggested that neural network-based modeling should be used to make correct management decisions in logistics systems, same as using managerial skills for innovation support [11].

The presented approach to modeling delivery processes is acceptable from the point of view of interests rationalizing of port operators responsible for logistics at transshipment terminals, as well as costs reducing of agricultural business for renting trucks for transporting grain to ports. Accurate forecasting allows the carrier to determine the required truck quantity for transporting grain to ports bypassing warehouses [12]. This will significantly reduce the cost of paying for grain transportation services by truck renters.

The issue of transport reliability has, at the same time, the highest priority in each delivery level of supply chains. It is because increasing transshipment volume through ports increases the probability of certain failures in multimodal transportation processes, especially in bulk goods delivery [13]. Products of agro-industrial manufacturing (AIM) are primarily mass shipments [14] for Ukraine, as an agrarian state.

Given that the ports of Ukraine export grain most than other agricultural products (according to data for 2018), we can talk about the relevance of the chosen research area. This makes it necessary to develop a forecast model considering modernity factors. A similar approach can be successfully applied when forecasting cargo flows during regular mass shipments.

Therefore, the research aim is to develop an approach that enables management decision-making to organize logistical issues at the port during interactions between two kinds of transport, considering forecasting results using evolutionary self-learning models.

2 Literature Review

Research study [15] describes the interaction problems of road and river transports in freight transportation. The example of cargo transshipment simulation from trucks to ships through the Boom Baru river pier (Indonesia, Palembang) proves the necessity for correct forecast models that predict volumes of cargo brought to the port by road immediately before of ship arriving.

Forwarding companies must have proven forecast transportation volumes obtained using self-learning systems in order to apply innovative technological solutions (direct grain overload). Forwarders can predict transshipment volumes with minimal error based on smart approaches. Different structures of neural and hybrid networks (based on a genetic algorithm) have shown their expediency as an excellent predictor [16] to find optional decisions of such problems.

The efficiency of ports functioning has been increased by using neuro-network mathematical tools in practice, considering the possibility of self-training and the ability to evolve [17–19]. Nevertheless, forecast models of this category have a drawback that does not allow them to be used with significant fluctuations in cargo flow with a small sample for the formation of the time series [20–22]. The volume increasing of grain transshipment through Ukrainian seaports is usually associated with an increase in the cost of this category of agricultural products in world markets. It explains big errors in classical predictors that are used for forecasting.

The design of initial databases (input signals) must be carried out, focusing on the principles of the fuzzy logic theory. It will ensure that a wide range of values, according to considered processes, are taken into account [23–25]. This nuance makes it possible to find many factors characterizing each value of the time series in order to reduce randomness in prediction.

The implementation of neural models makes it possible to abandon from the formation of a significant dataset when predicting cargo flows for medium terms. This nuance was established from the analysis of previous studies in this area [26, 27].

Article [28] describes patterns of possible oscillations for cargo flow arriving at transport hubs. Of particular importance in the development of cargo supply chains [29–31] is the interaction specificity of various categories of enterprises. The synchronization of logistics events [32] has a significant impact on the stability of supply chains, which guarantees the greatest synergy. This ensures the correct generation of management decisions based on the forecast values of cargo flows [33, 34].

It should be noted that there is a certain specificity that occurs when cereals are delivered to ports. It manifests itself primarily in the technological aspects of transportation. The railway is the main transport which delivery cargoes to Ukrainian seaports. However, a significant deficiency of specialized grain railway carriages has created a situation where road carriers carry out a long-distance delivery by trucks. The big transportation prime cost using trucks increases grain export prices, which negatively affects the competitiveness of domestic agricultural products in world markets.

On the other hand, using a large number of grain-trucks poses many technological problems, especially in terms of ensuring coordinated operation between two kinds of transport [35, 36]. It is firstly due to practical aspects of technical capabilities for grain transshipment according to the direct option of "automobile—ship."

3 Research Methodology

The reliability of supply chain functioning directly depends on the reliability of each element according to the studies carried out and also if the supply chain considers based on principles of systems theory. The negative impact can be reduced on the delivery process from possible failures of certain elements of the system during agricultural goods transportation. Following actions should be taken to achieve these improving:

1. Innovative models must be developed to predict future volumes of agricultural transportation. Process organization, in terms of technological aspects, will be improved using fuzzy-logic theory [37] and smart technology elements, such as a neural network. It will help to predict the volume of cargo transshipment through a port or using a blockchain system to design a flexible set of management decision-making in cargo transportation in Industry 4.0 [38] conditions, especially during an interaction between two kinds of transport in ports during transshipment according to the direct option.
2. The modern management decision support system is appropriate for flexible decision-making in the operational planning period and direct delivery of agricultural goods. This system can improve the quality of transportation management and must be designed according to the last tendencies in Industry 4.0. The best is using in this situation specialized software products, which can speed up processes of obtaining rational solution sets. Especially if the virtual complex is able not only to carry out certain calculations but also to carry out simulation of processes.

These two main actions will improve management issues in the transportation process of agricultural goods. Moreover, they also reduce the negative impact of possible delivery system failures, which will keep the reliability indicator in supply chains at required levels.

Mathematical tools of fuzzy-neural networks apparatus were chosen for prediction. Some elements of this specifical approach can be found in parallel researches [39–43].

The primary structure of the neural network was represented by seven inputs, which received signals about the corresponding values of transshipment volumes for each day of the week. A set of initial data is generated to identify the function $y(k)$ from statistical processing results. The original time series (1095 transshipment volume signals for the previous period) was divided in the next step into two identical halves (547 and 548 signals, respectively, in each part): the first 547 values used as input signals and the remaining 548 as output values. It is possible to approximately train the neural network in the first step using this research technology. Figure 2 shows changes in cargo transshipment volumes over time of network training.

The database used in the experiment consists of values set for training ($k = 1, \ldots, 274$) and a subset of the signals used for network testing ($k = 275, \ldots, 548$). It is possible to generate the number of inputs for future predictors using the ARX model due to such a previous separation of the source database. The best model structure can be found by retrieving the entire possible set of combinations $[m, n, d]$. The signals values $[m, n, d]$ varied randomly in the interval [1–110] during the ranking (Fig. 3).

A variant of the optimal predictive model based on ARX was found, which corresponds to the following quantitative structure: $[m, n, d] = [4, 4, 11]$. The proposed predictor ensures obtaining the value 8.3 of a square root of a mean square error (RMSE) on the training sample and on the test RMSE = 7.9.

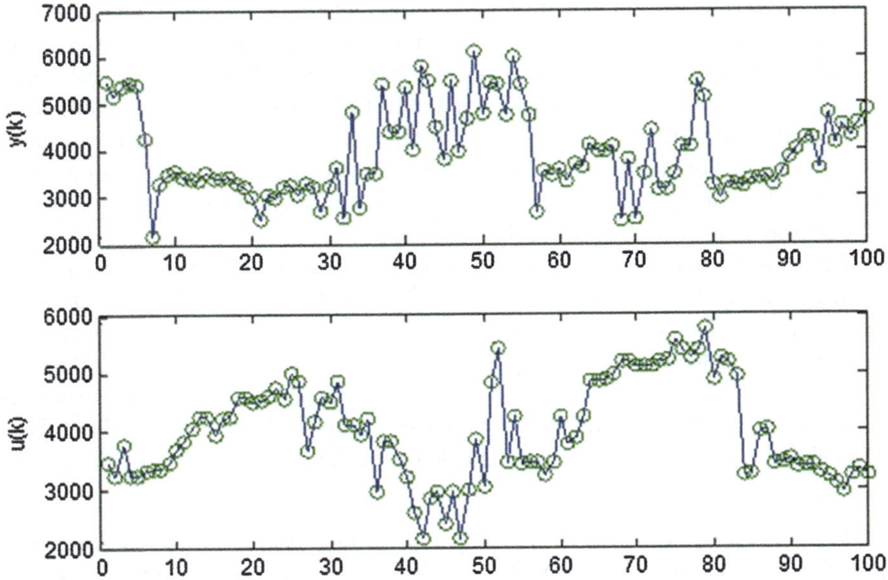

Fig. 2 Graphical interpretation of input signals (y) (grain transshipment volumes in the port) from output signals (u) of the same parameter

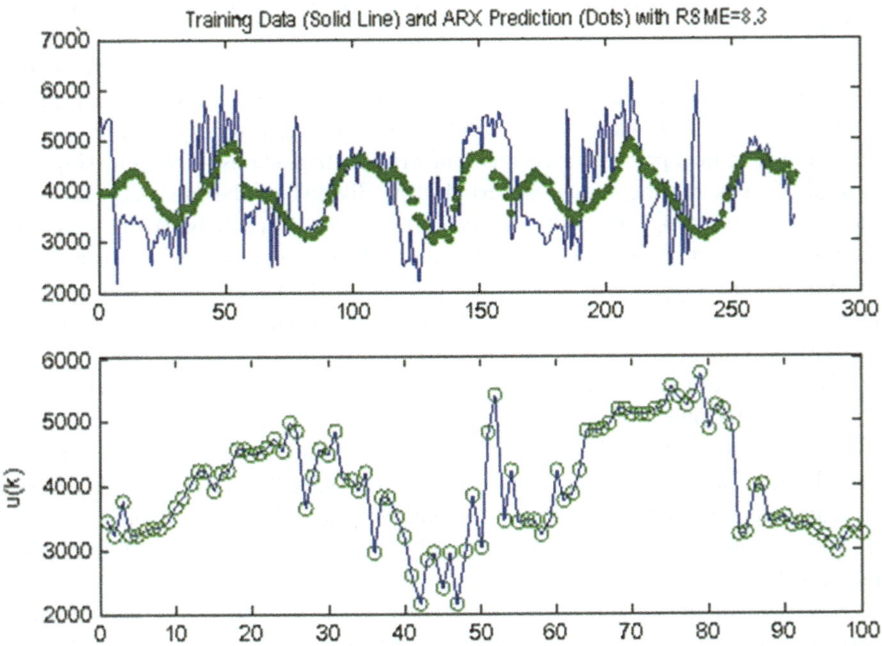

Fig. 3 Finding a rational structure for an ARX model

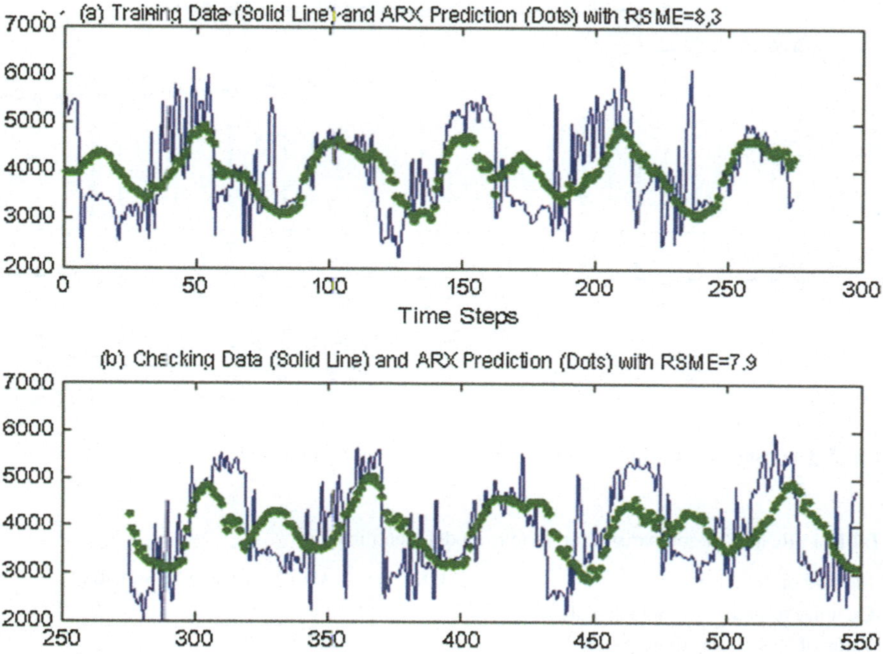

Fig. 4 Comparisons of simulation results between training samples (top graph) and checking (bottom graph) by ARX model

Graph (Fig. 4) shows the forecast trend obtained by the ARX model for predicting grain transshipment volumes at the port. Here, the blue curve denotes experimental data, and simulation results are represented by green dots.

A more accurate prediction can be obtained using a quasi-neural simulation. The best accuracy of identification is created using elements of fuzzy logic (fuzzy logical conclusion).

It includes elements of probability theory using a sequential forward search (SFS) approach. This simulation technology allows researchers to refine the input parameters of a future predictor. The model received an additional one variable at each experiment stage. This minimizes a mean square error of forecasting when applying SFS.

Figure 5 displays simulation results. Mean deviation values (forecast error) are highlighted in red circles when using a training sample. Deviation values are indicated by markers with green asterisks for a checking sample.

Table 1 presents the comparison results of different predictor identification methods. The ARX model is identified fastest but shows less accurate than others. At the same time, ANFIS using the SFS method for determining input signals has the best prediction accuracy, but a significant identification time. This is why the ARX model is useful for solving problems where the minimum time factor is important

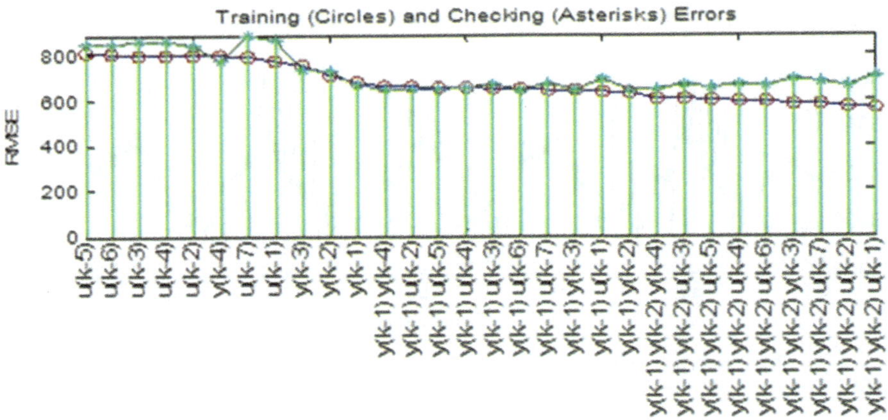

Fig. 5 Choosing input values of signals for a model using SFS method

Table 1 Results of comparison for various model identifications

Method of definition	ARX	ANFIS by using SFS method
Quantity of nodes for input signals	7	4
Value of RMSE for training sample	8.3	6.7
Value of RMSE for checking sample	7.9	5.3

for obtaining results. The ANFIS hybrid system is more suitable for predicting grain transshipment volumes since the error will be less than using heuristic models.

The number of input nodes for the ANFIS hybrid system has become less by three units (from 7 to 4) according to simulation results. That is why the first node receives a signal about volumes of grain transshipment on Monday and Tuesday; the second indicator inputs information about the values on Wednesday; third – data about volumes of cargo incoming on Thursday and Friday; information on values of cargo flows on weekends is supplied to the fourth node of the network.

Predictor error is 5.62% of empirical values after network training. The training period of the hybrid system does not exceed 15 eras, which in terms of machine time is 3.56 s. Figure 6 displays the visualization of ANFIS control using a training sample.

The designed ANFIS hybrid network makes it possible to generate 81 rules according to the previously described type of membership functions and the calculated number of nodes for supplying input signals. The optimal set of rules is determined using a genetic algorithm in this case.

In this case, the models become more adaptive to the appearance of sharp fluctuations in time series. After all, the hybrid network will be based not only on the local extremum but also considering a wider range of critical values of grain transshipment volumes in the port, which do not describe the general trend of time series.

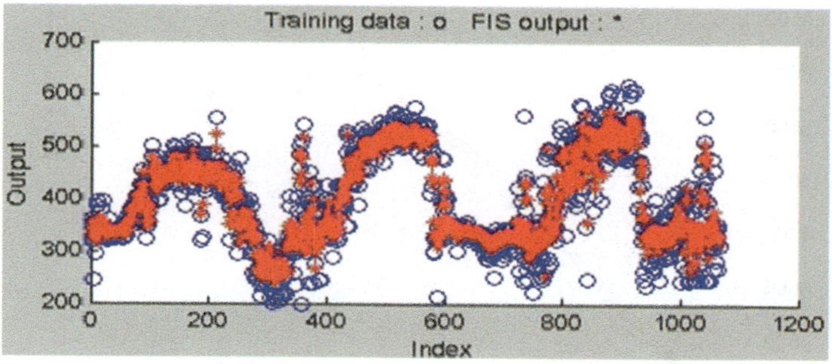

Fig. 6 Management of hybrid system ANFIS using a training sample

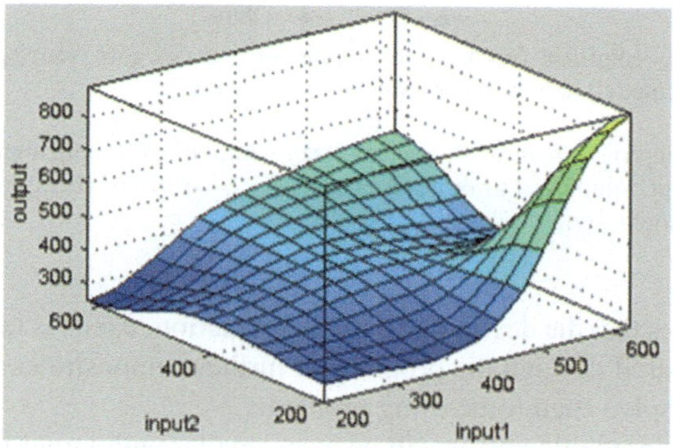

Notes: Input 1 – values of grain cargoes volume for training sample, ton*10; Input 2 – values of grain cargoes volume for training sample, ton*10; Output – results of forecasting, ton*10

Fig. 7 Dependence surface of forecasted results from incoming signals. Notes: Input 1—values of grain cargoes volume for training sample, ton*10; Input 2—values of grain cargoes volume for training sample, ton*10; Output—results of forecasting, ton*10

The hybrid network is trained based on a test sample to reduce prediction error. The trained ANFIS network showed an error of 4.49% as a result of checking it on the control sample.

The graphical dependence of input signals of volumes of grain transshipment in the port from the forecast results is shown in Fig. 7.

4 Results

The ANFIS hybrid system made it possible to predict seven values for potentially possible volumes of grain transshipment in the port zone. The forecast results made it possible to determine the error size in actual units (Fig. 8).

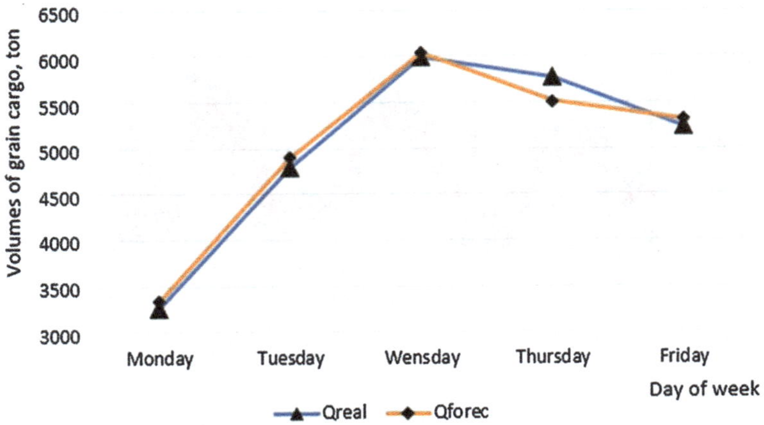

Notes: Q_{real}–empirical (real) input volumes of grain cargoes, ton;Q_{forec}– forecast (fuzzy-neural model) input volumes of grain cargoes, ton.

Fig. 8 Curves of empirical (real) and forecast (fuzzy-neural model) input signals. Notes: Q_{real}—empirical (real) input volumes of grain cargoes, ton; Q_{forec}—forecast (fuzzy-neural model) input volumes of grain cargoes, ton

Figure 8 proves that the deviation from the prediction results is less in actual units than when compared through the program. It further demonstrates the feasibility of using ANFIS to solve such forecasting problems.

Therefore, the prediction error will be even smaller with larger time-series sizes. It allows the researcher to better train and configure the network, especially if there are no significant fluctuations in the original database, which will make it possible to achieve an almost optimal result of forecasting.

4.1 Practical Aspects

The obtained forecast values allow planning the operation of overloading mechanisms and vehicles, which will carry grain to ports. Choosing transshipment mechanisms shall be at that time carried out based on technical parameters. Hourly capacity of the overloading mechanism must be first taken into account. This indicator will affect the following aspects of logistics processes:

- Sea ship service time (idle time under upload)
- Quantity of grain trucks required to provide combined work between vehicles and mechanization facilities
- Number of simultaneously used cargo mobile loaders
- Daily dispatch volumes, i.e., berth capacity

Mechanisms should possess a second specific feature. It is their mobility. This characteristic is particularly important when servicing several ships for a short time.

A trained neural network [44] was used to predict grain volumes that import into the port before the ship's arrival. The sample size for the experiment was six ships with a load capacity of up to 40,000 tons.

Predictive values are very important for making timely management decisions that must be consistent with Industry 4.0 policy. It is recommended to provide logistics nuances in the port when transferring cargo from one kind of transport to another based on three main factors related to port (terminal) capacities:

- Assessment of maximal port capacity in trucks at an hour according to technical features of infrastructure
- Definition of necessity port capacity in trucks at an hour based on coordinated interactions of vehicles and transshipment mechanisms
- Calculation of actual port capacity in trucks at hour based on results of cargo overload volumes forecast according to the fuzzy-neural model

4.2 *Assessment of Maximal Port Capacity in Trucks at an Hour According to Technical Features of Infrastructure*

The trucks' number is determined based on technical aspects according to terms of maximal quantities of vehicles that can accept by port zones per hour of operation. In this regard, the example of Mariupol seaport is one of the largest in quantities of trucks serviced. Port terminal can serve a maximum of 240-grain road-train trucks per day according to the maximum capacity. A comparison number of railway grain-carriages is 100 wagons in 24 h, which can be served by this port.

This study is ordered to determine the limitation of port capacity in management decision-making. It is assumed that trucks arrive at the port in an even flow. Grain overload occurs during an 8-h working day. Therefore, calculations of hourly capacities according to technical restrictions of the port are carried out by the next dependence:

$$\mathrm{CAP}_{\mathrm{tech}}^{\mathrm{hour}} = \frac{\mathrm{CAP}_{\mathrm{tech}}^{\mathrm{max}}}{T_{\mathrm{work}}^{\mathrm{port}}}, \tag{1}$$

where $\mathrm{CAP}_{\mathrm{tech}}^{\mathrm{max}}$—maximal port capacity in trucks at day according to technical aspects of loading infrastructures, trucks/day; $T_{\mathrm{work}}^{\mathrm{port}}$—operation time of transshipment mechanisms in the port during worker's duration, hour.

This condition shows that within an hour, no more than 30-grain road-train trucks can be serviced in the port. This nuance will be used in determining rational areas of management decisions regarding the organization of correct logistics in the port.

4.3 Definition of Necessity Port Capacity in Trucks at an Hour Based on Coordinated Interactions of Vehicles and Transshipment Mechanisms

Following condition must be met in order to ensure coordinated operation of transshipment mechanisms and vehicles:

$$\text{INTERVAL} = \text{OPERATION RHYTHM}, \tag{2}$$

where INTERVAL—trucks arrival time interval to the port, trucks/hour; OPERATION RHYTHM—operation rhythm of transshipment mechanisms, operation/hour.

It should be noted that two parameters play a key role in determining the required number of trucks to be serviced by the port. They are cargo capacity of grain road-train trucks and hourly operational productivity of overloading mechanisms. Mathematical dependency was derived to determine the required port capacity considering uninterrupted operation condition:

$$\text{CAP}^{\text{hour}}_{\text{neces}} = \frac{T^{\text{parking}}_{\text{1 h}} \cdot \text{TRANSCAP}^{\text{total}}_{\text{oper}}}{\text{LCAP}^{\text{average}}_{\text{truck}}}, \tag{3}$$

where $T^{\text{parking}}_{\text{1 h}}$—the time during which port achieve necessity volumes of grain cargoes for beginning to overload process, hour; $\text{TRANSCAP}^{\text{total}}_{\text{oper}}$—total cargo overloading capacity of transshipment mechanisms per hour, ton/hour; $\text{LCAP}^{\text{average}}_{\text{truck}}$—average loading capacity of trucks, ton.

The total cargo overloading capacity of transshipment mechanisms was calculated per hour by the following formula:

$$\text{TRANSCAP}^{\text{total}}_{\text{oper}} = \sum_{i=1}^{k} N_i^{\text{mech}} \cdot \text{TECHCAP}_i^{\text{mech}} \cdot K_i^{\text{ustime}}, \tag{4}$$

where N_i^{mech}—number of transshipment mechanisms of *i*th type, unit; $\text{TECHCAP}_i^{\text{mech}}$—technical cargo overloading capacity of transshipment mechanisms of *i*th type, ton/hour; K_i^{ustime}—coefficient of operation time using by *i*th type of transshipment mechanisms which load a ship, percent.

The distribution surface of the number of vehicles has been built based on calculations results of formulas (3) and (4) to ensure coordinated work between elements of grain supply chains to ports (Fig. 9).

Results of empirical observations of using mechanisms and trucks, which are operating for transshipment of grain cargoes to ships, were used in order to build the distribution surface of required port capacities in automobiles.

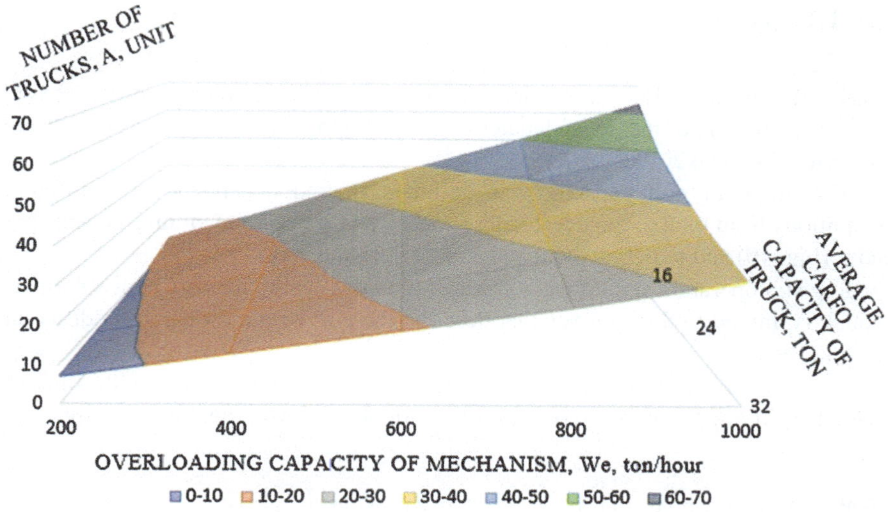

Fig. 9 Necessity port capacity in trucks at an hour based on coordinated interactions of vehicles and transshipment mechanisms

4.4 Calculation of Actual Port Capacity in Trucks at Hour Based on Results of Cargo Overload Volumes Forecast According to Fuzzy-Neural Model

The most important parameter is the actual port capacity in trucks. This factor helps design the flexible solutions at time management decision-making during cargo transshipment processes at the port. This indicator is determined by forecasting results (Fig. 8) derived from a trained fuzzy-neural network.

Conversion of classical dependency is used in this case to calculate actual port capacities to find the value of this indicator in trucks per hour. The final formula is presented as follows:

$$\mathrm{CAP}_{\text{fuzzy-neural}}^{\text{hour}} = \frac{Q_{\text{forec}}}{T_{\text{unload}}^{\text{idle}} \cdot \mathrm{TRANSCAP}_{\text{oper}}^{\text{total}} \cdot K_{\text{uneven}}^{\text{arrival}}}, \tag{5}$$

where Q_{forec}—forecast volumes got by using fuzzy-neural model, ton; $T_{\text{unload}}^{\text{idle}}$—truck idle time under unloading, hour; $K_{\text{uneven}}^{\text{arrival}}$—coefficient of vehicles uneven arrival at the port.

The coefficient of vehicles uneven arrival in the port was defined from observation of statistical data. It has an average value distributed in interval 1.1–1.8.

5 Discussion

The port capacity value was modeled in trucks per for different conditions based on the calculation of the dependencies (1), (3), and (5). The calculations results are presented in Table 2.

The graph of port capacity was designed to find dependency for different conditions from total cargo overloading capacity of transshipment mechanisms per hour (Fig. 10) and to define the management decision area.

Logistic operators should use a hatched area to develop management solutions, which is limited to three curves that describe the port operation based on different conditions.

System of making decisions on serviced trucks quantity in a specific time of reloading front operation can be represented by the following mathematically:

Table 2 Calculated results of port capacity according to various aspects

Maximal port capacity in trucks, trucks/hour	Necessity port capacity in trucks, trucks/hour	Actual port capacity in trucks, trucks/hour
30	13	28
30	25	33
30	38	34
30	50	26
30	63	23

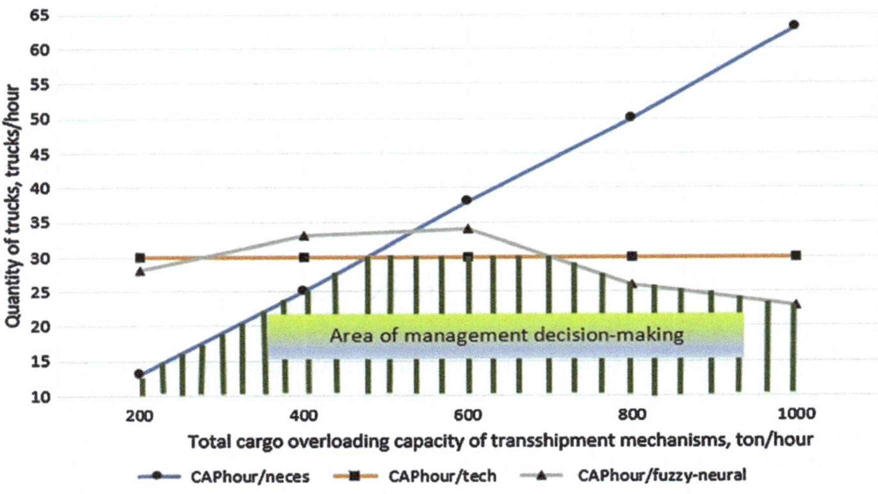

Fig. 10 Scheme for finding an area of management decision-making

$$\text{MAN DEC}\,(Q_{\text{trucks}}) = \min \begin{cases} \text{CAP}^{\text{hour}}_{\text{tech}}, & \text{technical condition}, \\ \text{CAP}^{\text{hour}}_{\text{neces}}, & \text{technological condition}, \\ \text{CAP}^{\text{hour}}_{\text{fuzzy-neural}}, & \text{actual condition}, \end{cases} \quad (6)$$

where MAN DEC(Q_{trucks})—management decision-making about logistics aspects of port functioning based on maximal potential quantities of trucks which can be serviced at the port during an hour, truck/hour.

Management decisions on this methodology allow port and carrier resources to be allocated in a manner that minimizes losses of time and financial resources. This approach is fully correlated with the Industry 4.0 concept on the appropriate allocation of resources to achieve the maximum possible profit while maintaining the high efficiency of manufacturing processes.

6 Conclusions

The methodology of management decision-making on logistics aspects organization in the port was presented based on the research results. The approach considers peculiarities of technical arrangements of overload fronts, technological aspects during the interaction between two kinds of transport, as well as prediction results obtained from the fuzzy-neural model. Obtained results of forecast about transshipment volumes have a lower average error (not more than 4.99%) due to using the ANFIS self-learning system than when using probabilistic methods.

The methodology presented advantages of management decision-making, as the actual calculation of truck quantities that will supply grain to the port is used. Therefore, planning with this indicator allows rational allocation of port resources, which is consistent with Industry 4.0 manufacturing policy.

The presented field of management solutions allows developing technological aspects in subsequent studies, in which logistics of port elements functioning will be the most optimal. It will, firstly, reduce ship service time, which will not exceed 2—3 days. It is in line with world standards and will improve supply chain efficiency in wholes.

The next step of the study will be the determination of the reasonable time of ships in port during loading considering management decisions-making based on the presented methodology.

The study will be aimed at reducing ship serviced time not only by organizing at a high level the interaction between two kinds of transport to ensure grain overload according to the direct option but also by determining the necessary grain reserves on port elevators, which should ensure the ship's loading in continuous mode.

Acknowledgment This research is partially funded by International Association for Technological Development and Innovations.

References

1. Ukrainian Ports. Analytic online reports. Available at: https://ports.com.ua/uk/analitics/gruzooborot-portov-2019-infografika
2. S. Zinchenko, V. Hvostovich, Improving methodology organizations of object repair transportation and technological system of the sea ports. Transp. Dev. **2**(3), 127–139 (2018)
3. I. Azarova, Complex Odessa Sea Port estimation on the sustainable development concept basis. Transp. Dev. **2**(3), 75–89 (2018)
4. D. Muzylyov, N. Shramenko, Mathematical model of reverse loading advisability for trucks considering idle times, in *New Technologies, Development and Application III. NT 2020*, Lecture Notes in Networks and Systems, ed. by I. Karabegović, vol. 128, (Springer, Cham, 2020), pp. 612–620. https://doi.org/10.1007/978-3-030-46817-0_71
5. M.L.R. Varela, G.D. Putnik, V.K. Manupati, G. Rajyalakshmi, J. Trojanowska, J. Machado, Integrated process planning and scheduling in networked manufacturing systems for I4.0: A review and framework proposal. Wirel. Netw. (2019). https://doi.org/10.1007/s11276-019-02082-8
6. S. Saniuk, A. Saniuk, D. Cagáňová, Cyber industry networks as an environment of the Industry 4.0 implementation. Wirel. Netw. (2019). https://doi.org/10.1007/s11276-019-02079-3
7. S. Haykin, *Neural Networks, A Comprehensive Foundation* (Macmillan College Publishing Company, New York, 1994)
8. J. Husar, L. Knapcikova, M. Balog, Implementation of material flow simulation as a learning tool, in *Advances in Design, Simulation and Manufacturing. DSMIE-2018*, Lecture Notes in Mechanical Engineering, ed. by V. Ivanov et al., (Springer, Cham, 2019), pp. 33–41. 10.1007%2F978-3-319-93587-4_4
9. N. Horňáková, L. Jurík, H. Hrablik Chovanová, D. Cagáňová, D. Babčanová, AHP method application in selection of appropriate material handling equipment in selected industrial enterprise. Wirel. Netw (2019). https://doi.org/10.1007/s11276-019-02050-2
10. A. Kotliar, Y. Basova, V. Ivanov, O. Murzabulatova, S. Vasyltsova, M. Litvynenko, O. Zinchenko, Ensuring the economic efficiency of enterprises by multi-criteria selection of the optimal manufacturing process. Manag. Prod. Eng. Rev. **11**(1), 52–61 (2020). https://doi.org/10.24425/mper.2020.132943
11. V. Švač, D. Cagáňová, Managerial skills for innovation support. Mobile Netw. Appl. (2020). https://doi.org/10.1007/s11036-020-01517-3
12. A. Kirichenko, Y. Lahmetkina, Multimodal container transportation of grain. World Transp. **13**(1), 166–125 (2015)
13. U. Delli, A.K. Sinha, Parallel computation framework for optimizing trailer routes in bulk transportation. J. Ind. Eng. Int. **15**, 487–497 (2019). https://doi.org/10.1007/s40092-019-0308-8
14. O.O. Gorshkova, Modernization of transport management system upon conveyance of bulk goods. Int. J. Emerg. Trends Eng. Res. **8**(3), 659–662 (2020). Available Online at: http://www.warse.org/IJETER/static/pdf/file/ijeter08832020.pdf). https://doi.org/10.30534/ijeter/2020/08832020
15. E. Buchari, D.D.A. Putranto, A. Rahman, Analysis of model loading and unloading time of ships at Boom Baru Port, Palembang, Indonesia, in *6th International Conference of Euro Asia Civil Engineering Forum*, MATEC Web of Conferences, vol. 138, (EDP Sciences, Les Ulis, 2017), pp. 1–9, 07012
16. G. Grebenişan, N. Salem, S. Bogdan, An approach of classification and parameters estimation, using neural network, for lubricant degradation diagnosis, in *Annual Session of Scientific Papers IMT ORADEA 2018*, MATEC Web of Conferences, vol. 184, (EDP Sciences, Les Ulis, 2018), pp. 1–6, 07009
17. K. Uyar, Ü. Ilhan, A. İlhan, Long term dry cargo freight rates forecasting by using recurrent fuzzy neural networks. Proc. Comput. Sci. **102**, 642–647 (2016)

18. N. Sari, W. Mahmudy, A. Wibawa, E. Sonalitha, Enabling external factors for inflation rate forecasting using fuzzy neural system. Int. J. Electr. Comput. Eng. **7**(5), 2746–2756 (2017)
19. I. López, M. López, G. Iglesias, Artificial neural networks applied to port operability assessment. Ocean Eng. **109**, 298–308 (2015)
20. E. Bal Beşikçi, O. Arslan, O. Turan, A.I. Ölçer, An artificial neural network based decision support system for energy efficient ship operations. Comput. Oper. Res. **66**, 393–401 (2016). https://doi.org/10.1016/j.cor.2015.04.004.
21. U. Gökkus, M.S. Yildirim, M.M. Aydin, Estimation of container traffic at seaports by using several soft computing methods: A case of Turkish seaports. Discrete Dyn. Nat. Soc. **2017**, 1–15 (2017). https://doi.org/10.1155/2017/2984853
22. V. Gosasang, W. Chandraprakaikul, S. Kiattisin, A comparison of traditional and neural networks forecasting techniques for container throughput at Bangkok Port. Asian J. Ship. Logist. **27**(3), 463–482 (2011). https://doi.org/10.1016/S2092-5212(11)80022-2
23. O. Pavlenko, N. Shramenko, D. Muzylyov, Logistics optimization of agricultural products supply to the European Union based on modeling by petri nets, in *New Technologies, Development and Application III. NT 2020*, Lecture Notes in Networks and Systems, ed. by I. Karabegović, vol. 128, (Springer, Cham, 2020), pp. 596–604. https://doi.org/10.1007/978-3-030-46817-0_69
24. S. Nie, Y. Zhong, M. Hu, Short-time traffic flow prediction method based on universal organic computing architecture, in *Proceedings of the 2012 2nd International Conference on Computer and Information Application (ICCIA 2012)*, (Atlantis Press, Paris, 2012), pp. 1226–1229
25. Z. Chen, T. Li, Port cargo throughput forecasting based on combination model, in *Proceedings of the 2016 Joint International Information Technology, Mechanical and Electronic Engineering*, (Atlantis Press, Paris, 2016), pp. 148–154. https://doi.org/10.2991/jimec-16.2016.25
26. V. Medvedev, V. Potemkin, Neural networks. MATLAB 6. Dialogue-MIFI, Moscow (2002)
27. N.A. Kizim, E.N. Yastremskaya, V.F. Senchucov, Neural networks: Theory and practice of application. Monograph. PH "INZHEK", Kharkiv (2006)
28. W. Wang, M. Baumann, X. Jiang (eds.), *Green, Smart and Connected Transportation Systems*, Lecture Notes in Electrical Engineering (Springer, Singapore, 2020). https://doi.org/10.1007/978-981-15-0644-4
29. N. Shramenko, D. Muzylyov, V. Shramenko, Methodology of costs assessment for customer transportation service of small perishable cargoes. Int. J. Business Perform. Manag. **21**(1/2), 132–148 (2020). https://doi.org/10.1504/IJBPM.2020.10027632
30. N. Shramenko, D. Muzylyov, V. Shramenko, Service costs in operational planning of transportation with small batches of cargo in city, in *Advances in Design, Simulation and Manufacturing III. DSMIE 2020*, Lecture Notes in Mechanical Engineering, ed. by V. Ivanov et al., (Springer, Cham, 2020), pp. 201–209. https://doi.org/10.1007/978-3-030-50794-7_20
31. V. Aulin, O. Pavlenko, D. Velikodnyy, O. Kalinichenko, A. Zielinska, A. Hrinkiv, V. Diychenko, V. Dzyura, Methodological approach to estimating the efficiency of the stock complex facing of transport and logistic centers in Ukraine, in *Proceedings Paper 1st International Scientific Conference on Current Problems of Transport. ICCPT 2019: Current Problems of Transport* (2019), pp. 120–132. https://doi.org/10.5281/zenodo.3387516
32. D. Muzylyov, N. Shramenko, V. Shramenko, Integrated business-criterion to choose a rational supply chain for perishable agricultural goods at automobile transportations. Int. J. Business Perform. Manag. **21**(1/2), 166–183 (2020). https://doi.org/10.1504/IJBPM.2020.10027634
33. A. Kumar, A. Kaur, M. Kaur, Fuzzy optimal solution of fuzzy transportation problems with transshipments, in *Rough Sets, Fuzzy Sets, Data Mining and Granular Computing. RSFDGrC 2011*, Lecture Notes in Computer Science, ed. by S. O. Kuznetsov, D. Ślęzak, D. H. Hepting, B. G. Mirkin, vol. 6743, (Springer, Berlin, 2011), pp. 167–170. https://doi.org/10.1007/978-3-642-21881-1_27
34. A. Kaur, J. Kacprzyk, A. Kumar, New improved methods for solving the fully fuzzy transshipment problems with parameters given as the lr flat fuzzy numbers, in *Fuzzy Transportation and Transshipment Problems*, Studies in Fuzziness and Soft Computing, vol. 385, (Springer, Cham, 2020), pp. 103–144. https://doi.org/10.1007/978-3-030-26676-9_6

35. J. Mehmann, F. Teuteberg, A conceptual framework of a decision support system for operational dispatching of agricultural bulk goods—An agent-based approach, in *Multiagent System Technologies. MATES 2014*, Lecture Notes in Computer Science, ed. by J. P. Müller, M. Weyrich, A. L. C. Bazzan, vol. 8732, (Springer, Cham, 2014), pp. 121–137. https://doi.org/10.1007/978-3-319-11584-9_9
36. N. Shramenko, D. Muzylyov, V. Shramenko, Model for choosing rational technology of containers transshipment in multimodal cargo delivery systems, in *New Technologies, Development and Application III. NT 2020*, Lecture Notes in Networks and Systems, ed. by I. Karabegović, vol. 128, (Springer, Cham, 2020), pp. 621–629. https://doi.org/10.1007/978-3-030-46817-0_72
37. A. Tryhuba, O. Bashynskyi, Y. Medvediev, S. Slobodian, D. Skorobogatov, Justification of models of changing project environment for harvesting grain, oilseed and legume crops. Indep. J. Manag. Prod. (Special Edition PDATU) **10**(7), 658–672 (2019). https://doi.org/10.14807/ijmp.v10i7.922
38. D. Muzylyov, N. Shramenko, Blockchain technology in transportation as a part of the efficiency in Industry 4.0 strategy, in *Advanced Manufacturing Processes. InterPartner-2019*, Lecture Notes in Mechanical Engineering, ed. by V. Tonkonogyi et al., (Springer, Cham, 2020), pp. 216–225. https://doi.org/10.1007/978-3-030-40724-7_22
39. H. Demuth, M. Beale, *Neural Network Toolbox for Use with MATLAB®* (The MathWorks Inc., Natick, MA, 2014)
40. B. Lei, D. de Ridder, D.M.J. Tax, F. van der Heijden, G. Xu, M. Feng, Y. Zou, *Classification, Parameter Estimation and State Estimation an Engineering Approach Using MATLAB*, 2nd edn. (John Wiley & Sons Ltd., Hoboken, NJ, 2017)
41. M.T. Hagan, H.B. Demuth, M.H. Beale, O. De Jesús, *Neural Network Design*, 2nd edn. (Ebook, 2014).
42. D. Simić, S. Simić, A review: Approach of fuzzy models applications in logistics, in *Computer Recognition Systems 4*, Advances in Intelligent and Soft Computing, ed. by R. Burduk, M. Kurzyński, M. Woźniak, A. Żołnierek, vol. 95, (Springer, Berlin, 2011), pp. 717–726. https://doi.org/10.1007/978-3-642-20320-6_73
43. Y. Ren, C. Wang, B. Li, et al., A genetic algorithm for fuzzy random and low-carbon integrated forward/reverse logistics network design. Neural Comput. Appl. **32**, 2005–2025 (2020). https://doi.org/10.1007/s00521-019-04340-4
44. N. Shramenko, D. Muzylyov, Forecasting of overloading volumes in transport systems based on the fuzzy-neural model, in *Advances in Design, Simulation and Manufacturing II. DSMIE-2019*, Lecture Notes in Mechanical Engineering, ed. by V. Ivanov et al., (Springer, Cham, 2020), pp. 311–320. https://doi.org/10.1007/978-3-030-22365-6_31

Index

D. Cagáňová et al. (eds.), *Advances in Industrial Internet of Things, Engineering and Management*, EAI/Springer Innovations in Communication and Computing,
https://doi.org/10.1007/978-3-030-69705-1

The manufacturer's authorised representative in the EU is Springer Nature Customer Service Centre GmbH, Europaplatz 3, 69115 Heidelberg, Germany. If you have any concerns regarding our products, please contact ProductSafety@springernature.com

Printed and bound by CPI Group (UK) Ltd, Croydon, CR0 4YY
30/09/2026
02225287-0004